Sonar and Underwater Sound

Sonar and Underwater Sound

Albert W. Cox
Raytheon Company

Lexington Books
D.C. Heath and Company
Lexington, Massachusetts
Toronto London

Cox, Albert W.
Sonar and underwater sound.

Includes index.
1. Sonar. 2. Underwater acoustics. 3. Anti-submarine warfare.
I. Title
VM480.C64 623.7'3 74-15547
ISBN 0-669-95935-9

Fourth printing, October 1988

Published simultaneously in Canada

Printed in the United States of America on acid-free paper

International Standard Book Number: 0-669-95935-9

Library of Congress Catalog Card Number: 74-15547

To Lillian

Contents

List of Figures

List of Tables

Foreword

Traditionally, the U.S. Navy has sought ways to apply the most advanced technology to the problems of defense. This has resulted in the proliferation of sophisticated sensors and weapons in the fleet. One of the most difficult tasks ahead for the Navy is integrating these complex systems to take advantage of their combined strengths while overcoming their individual limitations. Rather than simplifying the sailor's job, oftentimes technology has complicated it. From seaman to admiral, a detailed knowledge of these systems seems indispensable. Yet, they represent a range of scientific experience greater than any one man can master. The tendency, therefore, is to intellectually compartmentalize this knowledge; to think of radar, sonar and missiles as separate entities and to think of ships, aircraft and submarines as isolated platforms.

This is a logical approach and reduces the education problem to manageable size, but it prevents us from developing tactics which could benefit from symbiotic relationships. As the Navy grows smaller, fleet strength must be enhanced through a better integration of existing parts. The imaginative coupling of sensors, weapons and platforms can only be accomplished if officers and men understand the fundamental principles of each and appreciate the unique contribution each makes. The understanding of fundamentals is the cornerstone on which the ability to use knowledge rests.

Sonar and Underwater Sound directs itself to the fundamentals of sonar at the user level. In discussing not only the characteristics of the sonar but also the environment in which it operates and its integration in the anti-submarine warfare weapons system, it fills a basic need in the literature and should contribute significantly to the user's ability to envision new or better ways to use this versatile tool.

No officer of the line in the Navy, no matter what his sea going specialty may be, nor any civilian working on matters of naval strategy and tactics can afford not to understand the basics of sonar. Why and how sonars perform will dominate the tactics of every naval mission for years to come. To be a professional or even an amateur naval tactician, one must absorb the fundamental principles that Captain Cox has set forth in his book.

Stansfield Turner
Vice Admiral, U.S. Navy
Commander Second Fleet and
Commander Striking Fleet Atlantic

Preface

This book is the logical expansion of a paper entitled "Principles of Underwater Sound for the Tactical Officer" which I wrote in the fall of 1973 for use as a text during the study of tactics in naval warfare at the U.S. Naval War College, Newport, Rhode Island. The attempt in this book, as in the earlier paper, is to summarize those principles and concepts of underwater sound and sonar design parameters as they apply to the encounter at sea between the submarine and the anti-submarine units. The focus or viewpoint of the examination is that of the naval line officer in his role as tactical anti-submarine warfare commander at sea.

There can be little doubt that the rapid increase in the body of knowledge concerning underwater acoustics during the past three decades, and the translation of this knowledge into sonar system technology, requires, on the part of the user line officer, a greater depth of perception and understanding of the basic principles involved than heretofore considered necessary. This book is an attempt to present these principles in the context of anti-submarine warfare operations as the minimum technical baseline deemed necessary. Based on the premise that those having the greatest need for such a text are either about-to-be or newly commissioned officers, or those officers whose formal education seems or is long ago, the material is presented generally at the undergraduate level. It is the author's hope that the book will be useful not only as a baseline academic text, but as a convenient refresher handbook during subsequent assignments involving sonar systems and anti-submarine warfare.

In the approach to the subject matter, the book is structured around the sonar equation—the basis for quantitative expression of sonar system detection performance. Four perspectives or points of focus are addressed. The first point of focus is the nature of sound, its behavior in the ocean environment, and the use of the sonar equations to predict sonar detection performance. The second point of focus is the role of the sonar in the anti-submarine warfare system. The third point of focus is on some sonar system design concepts and their interaction with the acoustic geometry of sound transmission in the ocean and the tactical geometry of the encounter with the submarine. The final point of focus is the acoustic counter-detectability of own anti-submarine units by opposing submarines and its impact on the ASW encounter.

It is a pleasure to acknowledge the support and assistance of a number of people, without whose support and encouragement this book would not have been possible. First among these are my associates in the Engineering Laboratory at the Raytheon Company, Submarine Signal Division, for many excellent technical discussions and criticisms. Next are the War College Staff and the many students of the 1973-74 class who were kind enough to comment on or criticize my prior work. More specifically, I wish to thank Rear Admiral S.L. Gravely, Jr., USN, Commander Cruiser Destroyer Group TWO: Dr. D.M. Viccione of the New London Laboratory, U.S. Naval Undersea Systems Center, and Cdr. H.C. Huber, USN (Ret) of the Naval Air Systems Command for their constructive criticisms.

1 Introduction

Principal Functions of Sonar

Sound energy travels in sea water at a speed of 1500 meters/second plus or minus about 3% depending upon temperature, salinity, and pressure. It can be generated, controlled, directed, transmitted, and received to perform in hydrospace (water) all those functions (communications, navigation, detection, tracking, classification, etc.) which are accomplished in aerospace with electromagnetic energy. Sound energy travels much better in sea water than it does in the air. Under ideal conditions, underwater sound signals have been transmitted over distances of hundreds and even thousands of miles. However, the sea as a medium for sound propagation is extremely variable so that ideal conditions do not prevail generally and it is essential for the user of the sonar equipment to understand the conditions in effect at the time of operation of the sonar equipment so as to get a measure of the performance to be expected at that particular time. The main factors affecting sonar performance are propagation conditions and own ship's noise. These will be discussed in some detail further herein.

Sonar is used for various purposes in naval, industrial, and oceanological applications. With man's increased concern about his environment, including a growing interest in the oceans of the world, the non-military uses of sonar have increased markedly during the past twenty years. An awareness of these uses may be obtained from the listings in Tables 1-1 and 1-2.

The naval applications of sonar encompass a variety of equipments for use on various platforms to perform the functions listed in Tables 1-3 and 1-4.

Passive and Active Sonar Types

A passive sonar system is basically an acoustic sound receiving system mostly used to listen to sounds produced by submarines and

Table 1-1
Industrial Sonar Applications

Fish Finder/Fish Herding
Oil and Mineral Exploration
River Flowmeter
Echo Depth Sounder
Acoustic Navigation Beacon
Acoustic Marine Speedometer/Doppler Navigator
Acoustic Telemetry and Control
Swimmer's Telephone
Acoustic Holography
(Ultra) Sonic Cleaner
Viscosimeter
Ultrasonic Grinding/Drilling
Process Measurement/Control
Anti-stranding Depth Sounder
Acoustic Ship Docking System

Table 1-2
Oceanological Sonar Application

Depth Sounding
Subbottom Geologic Mapping
Ocean Topography
Acoustic Path Stability Measurement
Bathyvelocimeter
Emergency Telephone
Seismic Simulation and Measurement
Biological Signal and Noise Measurement
Sonar Calibration

Table 1-3
Principal Functions of Naval Sonar

Detection	Deciding that a target is or is not present.
Classification	Deciding that a detected target does or does not belong to a specific category (a submarine, for example).
Localization	Measuring at least one of the instantaneous position and velocity components of a target (either relative or absolute), such as range, bearing, range rate, or bearing rate.
Navigation	Determining, controlling, and/or steering a course through a medium; includes avoidance of obstacles and the boundaries of the medium.
Communication	Transmitting and receiving acoustic power and information.
Countermeasures	Acting to neutralize or oppose the plan of another.

Table 1-4
Typical Navy Sonar Platforms

Surface Ships	Torpedoes
Submarines	Mines
Fixed Wing Aircraft	Fixed Position Systems
Rotary Wing Aircraft	

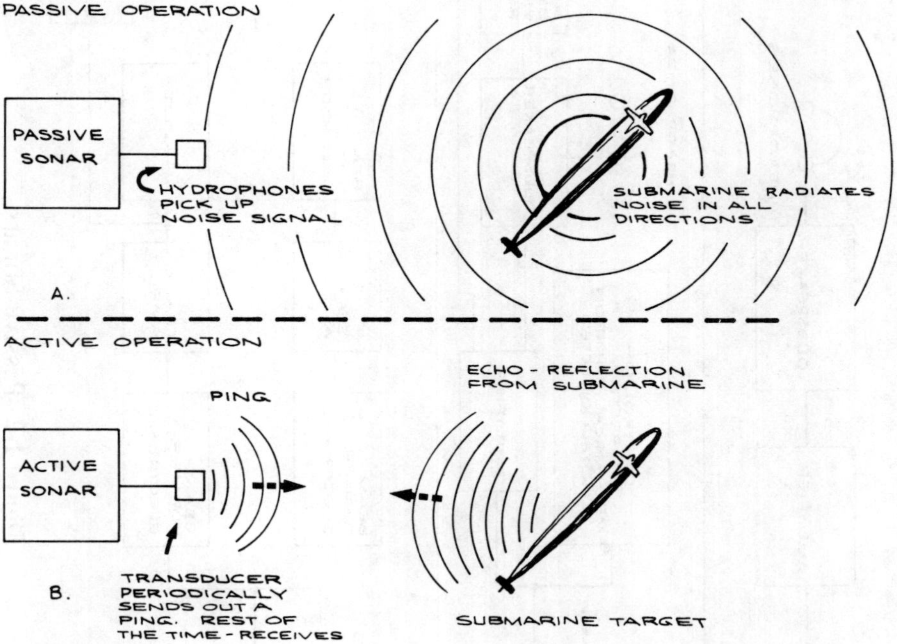

Figure 1-1. Passive and Active Operation

other vessels. It does not put any acoustic energy into the water.

An active sonar system is based on the echo-ranging principle where an acoustic signal is radiated and returning echoes from targets are received, similar to the function of a radar set.

Passive and active operation are illustrated in figure 1-1. Figures 1-2 and 1-3 provide simplified block diagrams of such rudimentary systems. Most active systems can also listen but only over the narrow frequency band used for echo returns. Advanced sonar systems now in use have both passive and active capabilities with selection of operating modes available to the operator.

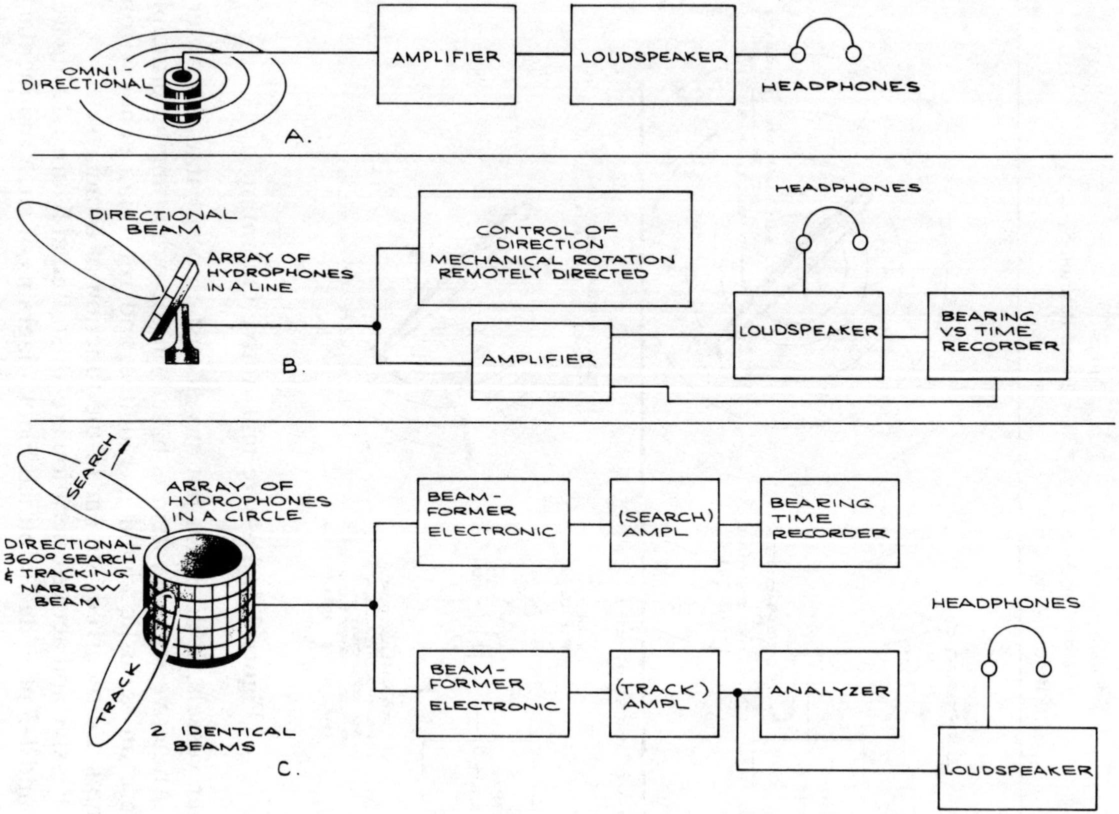

Figure 1-2. Passive Sonars

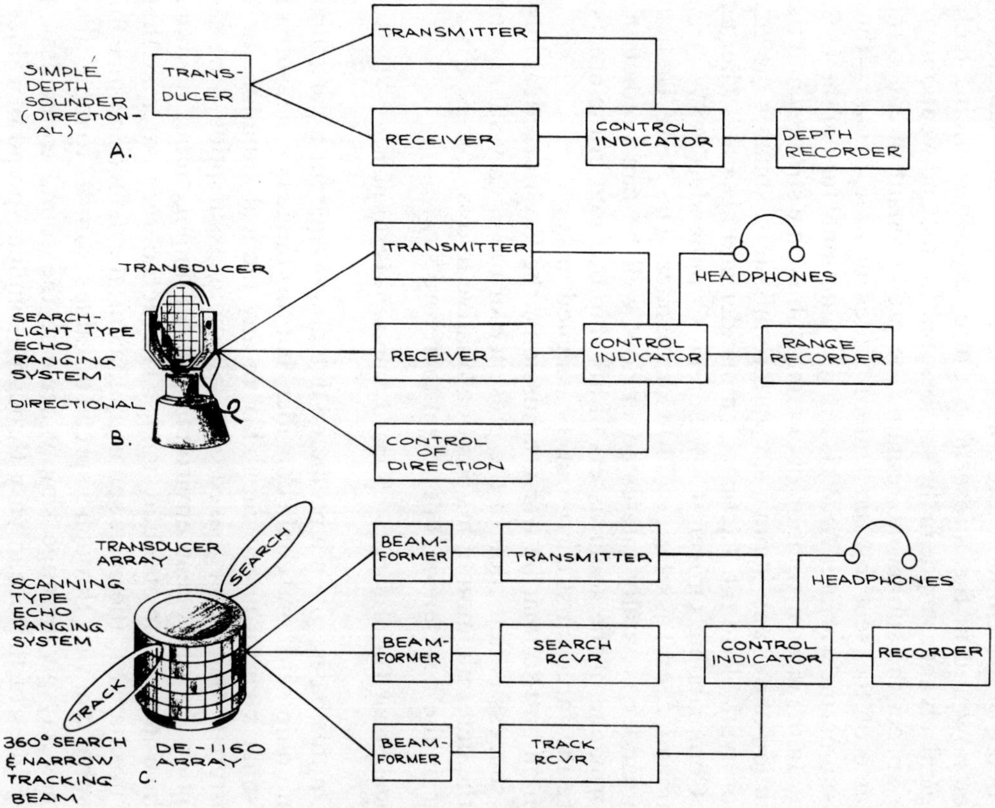

Figure 1-3. Active Sonars

Tactical Sonar Performance

As a major world power and insular nation, the use of the world seas is vital to the continued preservation of the security and economic well being of the United States, as well as to the world balance of power now and in the foreseeable future. Notwithstanding technological advances made in the sensors, platforms, and weaponry of outer space, the submarine remains the predominant threat to the free use of the seas by the United States in time of war—thus anti-submarine warfare (ASW) is a major functional task of the U.S. Navy. In the implementation of our anti-submarine strategy and the execution of our anti-submarine warfare missions, sonar is the key sensor available to the Tactical Commander and his staff, whether ship Commanding Officer (CO), Screen Commander, ASW Area Commander, or Task Force (TF) Commander. The ASW Tactical Commander is vitally interested in how well he can predict the performance of the sonar assets of his ship or force in the area(s) where the force must carry out its assigned mission.

In all anti-submarine warfare missions, the key to tactical mission success is detection of the presence of the enemy submarine and the determination of his position relative to own forces. Given sonar as the key sensor, one might therefore say

Tactical Sonar Performance = Sonar Equipment Performance (1.1)

On this basis, given a predicted detection range for a particular sonar equipment, one could consider that this range is a measure of the sonar's performance capability in any tactical situation. It is unfortunate that this is based on a fallacious assumption—the assumption that the ocean environment is an homogeneous medium for the transmission of sound signals. The facts of the matter are that with no change in the basic sonar equipment, the detection range for this sonar can vary through a spread of values from 1/2 to more than 6 times any given range simply because the sound transmission capabilities of the ocean vary through a similiar spread of values. Each sonar is designed to detect the arrival of a certain minimum sound energy level arriving at the sonar. The distance or range over which the sound energy traveled to produce this minimum detectable signal level at the sonar hydrophones is dependent upon how much energy was lost en route. Some parts of the ocean transmit sound more readily than others. In addition, as a sound transmission medium, the same geographical place in the ocean will vary in its sound transmission ability from season to season, and even from hour to hour on the same day. Because of this variability in the ocean, it is apparent that detection range is ocean dependent as well as sonar equipment dependent, and that tactical anti-submarine mission success depends not only on the sonar equipment, but on the ocean where the mission is conducted. Formula 1.1 is therefore restated:

$$\text{Tactical Sonar Performance} = \text{Sonar Equipment Performance} + \text{Ocean Transmission Performance} \quad (1.2)$$

A major purpose of the discussion to this point is to focus on the dual dependency of tactical sonar performance on both sonar equipment and the ocean. It was stated earlier that sonar equipment is designed to detect the arrival of some defined minimum level of sound signal energy. The sonar having thus been specified and subsequently built, it can be surmised that its design characteristics can be measured and expressed quantitatively. Unlike the sonar, the other variable in our problem is the ocean, which occurs in nature, and which the ASW Commander must take as it is. Sufficient measurements of appropriate ocean parameters have been made to permit some degree of stating and predicting the sound transmission characteristics of the various oceans as a function of season. These measurements coupled with a body of theory on sound transmission in the ocean provide a quantitative basis for expressing the transmission characteristics of the tactical commander's ocean. The expected tactical performance of the sonar, i.e., detection range in a particular ocean area, can be computed by combining the quantitative expressions of sonar design performance and ocean transmission characteristics in an equation called the *Sonar Equation.* This equation will be discussed in detail in Chapter 3.

At this point, it should be apparent that before the ASW sonar assets of a tactical ASW Commander are deployed, it would be well to relate the sonar equipment performance expected to the sound transmission characteristics of the sea expected in the area of operation. The next two chapters discuss the following as the prelude essential to effective ASW planning and operations:

1. Behavior of sound in water.
2. Variability of the ocean environment.

3. Measuring the ocean environment.
4. The Sonar Equation.

2 Sound and the Ocean Environment

The Nature of Sound and its Behavior in Water

From the ASW (or sonar) point of view, sound may be usefully conceived in several ways. First is the concept of sound as a form of energy. As such, it is subject to those laws of physics dealing with conservation of energy, foremost of which is the notion that the quantity of energy in the universe is constant and may not be destroyed but only translated or changed in form. Second is the concept that three elements are essential for the transmission and detection of sound or acoustic energy—a sound source, a medium, and a detection. Third is the concept that acoustic energy is in fact a form of mechanical energy.

To illustrate, assume that a diaphragm immersed in water is caused to vibrate by either electrical or mechanical means. The vibration of the diaphragm in turn causes a disturbance in the water medium around it. In other words, the mechanical movement (energy) of the diaphragm has been translated to the water medium. If a second diaphragm was placed in the water and the disturbed water caused it to vibrate, the acoustic energy would have been translated to this second diaphragm. The vibrations of the second diaphragm can be measured by electrical means to complete the flow of mechanical energy from the first diaphragm to the second. This second diaphragm is a detector of mechanical energy.

If the quantity of energy at the vibrating source and that at the detection had been measured, it is likely that the value at the detection would be less than that at the source. Some of the energy imparted to the medium would have been imparted to molecules of water in the medium. This added energy within each molecule results in increased motion of the molecule and can be considered as "heat," analogous to friction in other mechanical applications.

Sound waves are translated through the medium by the movement of the particles of the medium. Consider again the vibrating sound source diaphragm. As it is caused to move in one direction, it

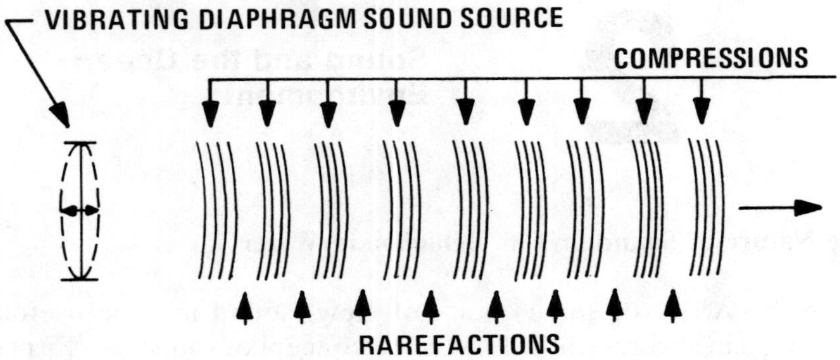

Figure 2-1. Basic Longitudinal Wave Concept

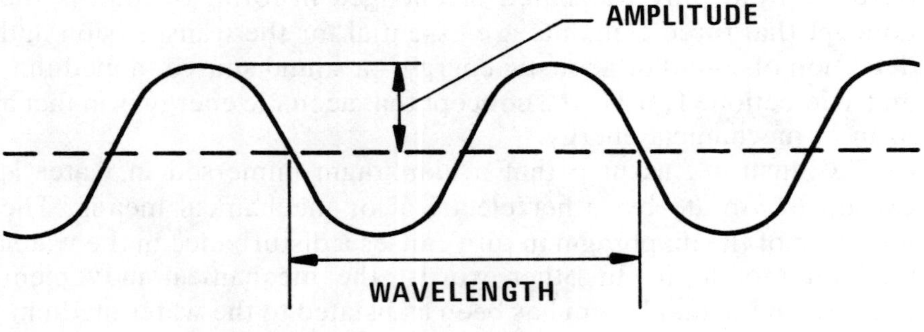

Figure 2-2. Vertical Cross-section of Longitudinal Waves

tends to push or compress the particles of the water medium in that direction. If the diaphragm is now caused to move in the direction opposite to its first motion, it relaxes the pressure against the medium in the direction of original movement. As the diaphragm continues to move to and fro, it can be seen that the particles of the medium are alternately compressed and relaxed.

Considering now the particles of the medium thus set in motion, the particles originally disturbed by the diaphragm in turn disturb particles adjacent to them and the energy wave thus travels through the medium. This is illustrated by what happens when a rock is dropped in a pond. A wave motion is observed moving outward from the point of entry of the rock. The crests of the waves are compressions; the valleys between waves are areas of low pressure where

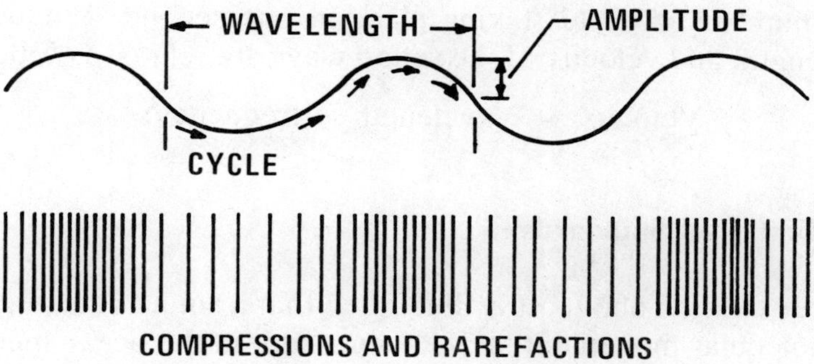

Figure 2-3. Composite View of Longitudinal Waves

the particles of the medium have moved apart after having been compressed. These low pressure areas are called rarefactions.

Sound or acoustic energy travels through the water medium in compression or longitudinal waves as described above—waves in which the particles of the medium move back and forth in the direction of the wave motution. A longitudinal wave as discussed above is depicted graphically in Figures 2-1 and 2-2.

Longitudinal waves have characteristics other than direction, compression, and rarefaction. Consider again the water waves resulting from the rock in the pond. These waves have visible alternate crests and valleys, as depicted in the simple wave diagram in figure 2-2.

These additional characteristics of the longitudinal waves are amplitude and wavelength, as shown in figure 2-2. By combining the sketches of the longitudinal wave and the simple wave, a wave of sound energy in water is portrayed (figure 2-3).

It can be seen in figure 2-3 that one complete cycle of a longitudinal wave moves a particle from rest position to maximum displacement in one direction, then to maximum displacement in the opposite direction, and finally back to the rest position. The wavelength is the distance the disturbance moves during one cycle or period of vibration. Other quantities of interest are amplitude and frequency. The amplitude of the sound energy wave is the distance from rest position to the maximum displacement from rest. Amplitude is a measure of the intensity of the sound energy transmitted. Frequency is the number of cycles taking place in one second. Frequency, wavelength, and velocity of the sound wave are related as follows:

$$\text{Velocity} = \text{Wavelength} \times \text{frequency}. \qquad (2.1)$$

Basic Units and Standards

In the discussion above, it is illustrated that a sound source generates molecular movement and thus pressure in the ocean medium, and that the amplitude of the sound waves or pressure thus generated is proportional to the pressure exerted on the medium by the source. From this, it follows that the sound pressure level (SPL) generated is a suitable measure of the quantity of sound in the water. Sound pressure levels are commonly expressed in units of decibels (dB). The decibel system was selected by acousticians for a number of logical reasons. First, it is a logarithmic system, which is convenient for dealing with large changes in quantities. It also simplifies computations, since multiplication and division are reduced to addition or subtraction, respectively. Second, human senses have an approximate logarithmic response to stimuli such as light, sound, and heat. For example, the human ear perceives about the same change in loudness between 1 and 10 units of pressure as it perceives between 10 and 100 units of pressure. And finally, in the area of underwater acoustics, the primary interest is in ratios of power levels and signal levels rather than absolute numerical values. In the decibel system, the bel is the fundamental division of a logarithmic scale for expressing the ratio of two amounts of power. The number of bels to express such a ratio is the logarithm to the base 10 of the ratio, written $\log_{10}$ (Ratio).

If P_1 is the amount of power being considered, and P_0 is a reference quantity of power, and N is the number of bels denoting the ratio P_1/P_0 then

$$N = \log_{10} P_1/P_0 \text{ bels.}^{a} \qquad (2.2)$$

If P_1 is 10 times P_0, then $N = \log_{10} 10$, or 1 bel.

Acousticians decided the bel was a unit too large for application in their field, and subsequently adopted the decibel (1/10 bel) as their

[a] $\log_{10} P_1/P_0$ is read "the logarithm to the base 10 of the ratio P_1/P_0."

basic logarithmic unit. In decibel units, the expression above then becomes:

$$N = 10 \log_{10} P_1/P_0 \text{ decibels (dB).} \tag{2.3}$$

If P_1 is 10 times the value of P_0, then

$$N = 10 \log_{10} 10, \text{ or } 10 \text{ dB.}$$[b]

In sonar computations, one is interested in comparing amounts of power radiated from two sonar transducers, and also in comparing the sound pressure levels resulting from the radiation of the respective transducers. As previously stated, expression 2.3 is the appropriate relationship for comparing power levels. It can be shown that the corresponding expression for comparing sound pressure levels in decibels is

$$N = 20 \log_{10} P_1/P_0 \text{ (dB).}$$

Since by definition, a decibel is an expression of the ratio of a pressure or intensity level to a standard pressure or intensity level, the reference used for comparison must always be given for the term to be meaningful. Until comparatively recently, the reference standard commonly used for acoustic pressure in water for sonar applications is the microbar (μb or μbar), which is 1 dyne/cm^2. A dyne/cm^2 is defined as the amount of force required to cause a mass of one gram to alter its speed by one centimeter (cm) per second for each second during which the force acts. The quantity 20 dB re 1 μbar is read "20 dB with reference to 1 μbar. The quantity 20 dB// 1 μbar is read the same, *re* and // having the same meaning. Sometimes the quantity might be expressed 20 dB// 1 dyne/cm^2. Since 1 dyne/cm^2 = 1 μbar, the reference quantity is the same.

Use of the microbar reference frequently results in negative values of power or sound pressure levels. In recent years, the use of the micropascal (μPa) as a reference quantity has become more widespread. One micropascal is equal to one micronewton (μN) per square meter (m^2). One micronewton per square meter (1 μN/M^2) is defined as the force required to cause a mass of one kilogram to alter its speed by one meter per second for each second which the force

[b]Unless specifically stated otherwise, logarithmic expressions used hereafter in this text will be understood to be the common logarithm to the base 10, and the base 10 notation will be omitted, i.e., 10 log 10 is read the same as 10 $\log_{10}$ 10, 10 times the logarithm to the base 10 of 10, where P_1 is the quantity of pressure of interest, and P_0 is the reference pressure.

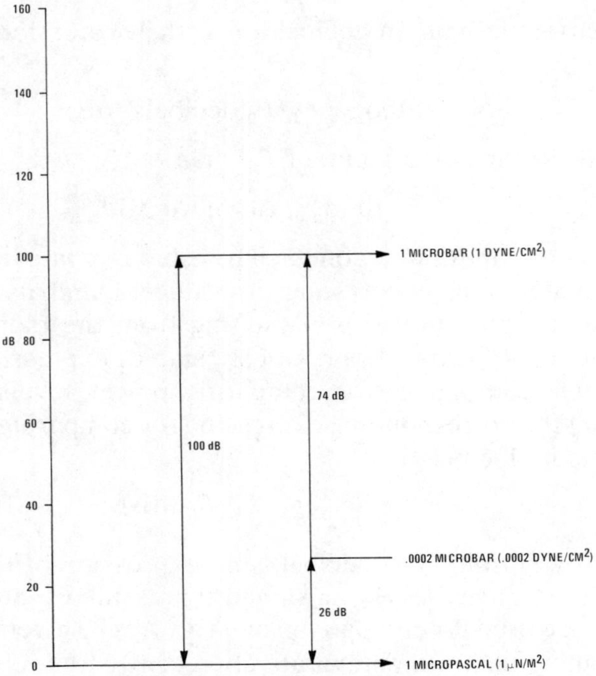

Figure 2-4. Reference Pressure Levels

acts. The relationship of the micropascal to the microbar is as follows: in units of force,

$$1\ \mu Pa = 10^{-5}\ \mu\text{bar} = \frac{1}{10{,}000}\ \mu\text{bar}.$$

In logarithmic quantities,

$$100\ \mu Pa = 1\ \mu\text{bar}.$$

Using the latter expression, we can depict graphically the relationships between these two reference standards in Figure 2-4 below.

To complete the picture of reference quantitites, there is a third reference standard used in acoustic work in air. For this work, a zero dB reference level of .0002 dynes/cm² was selected because through experimentation it was found to be the least audible sound level detectable by the human ear. This reference quantity is also shown in Figure 2-4.

To be precise and completely unambiguous, some writers will

use the notation "//(1 μPa) $(\text{Hz})^{\frac{1}{2}}$" to indicate that the value is a sound pressure spectrum level.

Since both the microbar and the micropascal are in use as reference standards in the literature today, the reader needs an awareness of their differences, and the capability to readily convert from one to the other is desirable. Table 2-1 contains some equivalent quantities expressed to each reference.

Table 2-1
Comparative Values Referred to Different Reference Standards

Reference quantity	*Comparative values (dB)*						
1 microbar (1 μbar)	100	36	−10	10	−42	16	−51
1 micropascal (1 μPa)	200	136	90	110	58	116	49

Table 2-2 is a similar table, but only partially completed. It is suggested the reader complete this table. The correct values are shown in Table 2-3.

Table 2-2
Comparative Values Referred to Difference Reference Quantities

Reference Quantity	*Comparative Values (dB)*						
1 microbar (1 μbar)	130	−41		−25		12.5	
1 micropascal (1 μPa)			113	75	98.5	112.5	160

For purposes of most calculations made for the purpose of expressing sonar performance, certain quantities of interest usually are specified as the sound pressure levels in a frequency band 1 hertz (Hz) wide. Values thus expressed are called spectrum levels.

Table 2-3
Completed Version of Table 2-2

Reference Quantity	*Comparative Values (dB)*						
1 microbar (1 μbar)	130	−41	13	−25	−1.5	12.5	60
1 micropascal (1 μPa)	230	59	113	75	98.5	112.5	160

The intensity levels of some common sounds as well as some of those of interest in ASW are shown in figure 2-5.

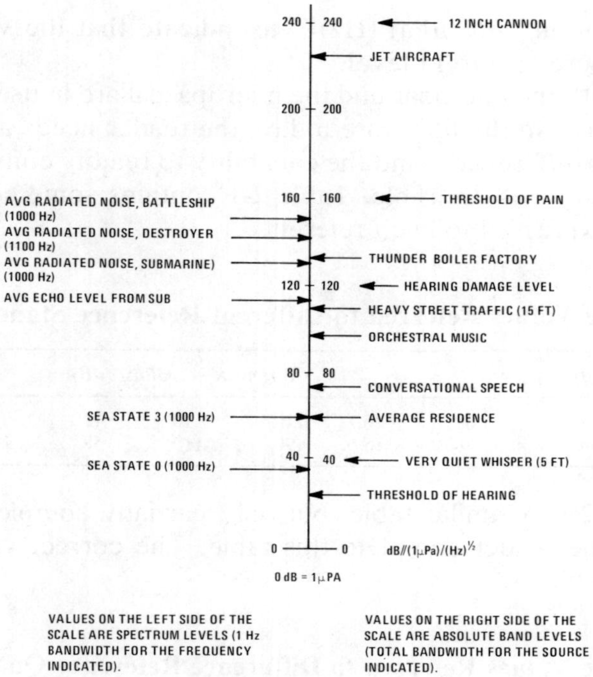

Figure 2-5. Nominal Values of Some Common Sounds and Some Sounds of Interest in ASW

Transmission of Sound in the Ocean Environment

Based on the foregoing review of the theory of sound transmission in water, it is now appropriate to take a more detailed look at the sound transmission characteristics of the oceans in which the ASW Tactical Commander will plan and execute his ASW missions. From the point of view of sound transmission, the ocean is not the homogenous medium it might appear to be. In addition to foreign materials which restrict sound travel, it has boundaries (the bottom and the surface) which influence sound travel, and as indicated previously, it absorbs some of the sound energy of interest in the form of heat. These characteristics are factors which impede, divert, or diminish sound energy and which delineate the sound transmission paths available for use to the ASW Commander and are the next topics for consideration. In the discussion of these characteristics, the reader should keep in mind those statements and illustrations cited above relating to the conservation, translation, and transformation of energy.

Consider a sound source in the ocean, emanating a sound energy of interest, and a sound energy detector some distance from the source. Sound energy en route to the detector from the source is diminished by any one or combinations of the following.

Spherical Spreading

From a source, sound can be considered to be projected outward as waves in the form of concentric spheres, similar to the waves generated by a stone dropped into a pool. The reduction in acoustic intensity due to an increase in the area over which a given acoustic energy is distributed is said to result from spreading and is called spreading loss. Spherical spreading loss is calculated as follows:

$$\text{Spherical Spreading Loss (dB)} = 20 \log R, \tag{2.5}$$

where R is the range in yards.

Absorption

During the propagation process of compression and rarefaction, the sound wave loses a certain amount of energy to each molecule in the water. The added energy within each molecule results in the increased motion of the molecule and can be considered as "heat." The amount of energy lost will depend on the frequency of the sound; as the frequency goes up, the losses go up. We cannot use (detect) this heat, so it is lost for our purposes. Absorption is expressed in decibels per unit distance. While absorption varies with a number of factors such as salinity and scattering, a dominant factor is frequency. Absorption Loss is calculated as follows:

$$\text{Absorption Loss (dB)} = aR \times 10^{-3},$$

where R = range in yards and

a = absorption coefficient in dB/kYd.

The relative effects of each of these factors on the velocity of sound is shown in the following expression:

$$C = 4422 + 11.25t + .0182d + S - 34,$$

where C = sound velocity in feet/sec.

t = ocean temperature in degrees F.

d = depth in ft., and

S = salinity in parts per thousand (ppt).

At very high frequencies (above 500 kHz) the viscosity of the ocean medium is the dominant source of absorption. In the 5 to 500 kHz frequency range, absorption is a function of some of the properties of the magnesium sulfate (M_8SO_4) molecule—one of the minor dissolved salts in the ocean. At frequencies below 5 kHz, for reasons yet undetermined, absorption has been found to vary at a rate different from those stated above. At a fixed temperature of 39°F., because of these factors, the absorption coefficient varies approximately with frequency in the following manner:

$$a = \frac{0.1f^2}{1 + f^2} + \frac{40f^2}{4100 + f^2} 5 \times 10^{-4}f^2, \qquad (2.6)$$

where a is in dB per kiloyard of range and

f is frequency in kHz.

In this expression, the first term corresponds to the frequency region below 5 kHz, the middle term the 5-500 kHz region, and the third term the region above 500 kHz. Some typical absorption coefficients calculated from this expression are shown in Figure 2-6.

Scattering and Reverberation

While we tend to think of the sea as basically homogenous, both within and on its boundaries, a variety of "foreign" materials exist in the sea. The boundaries are the ocean bottom and the surface. These boundaries vary in irregularity from location to location and from sea state to sea state. Within the boundaries of the sea, there are other foreign materials such as dust, sea organisms, fish, etc. When a sound wave strikes any of these types of foreign or non-homogenous items, some of that sound is reflected and re-radiated and re-reflected and re-reradiated. The effect is the same as if one sounded a short sound pulse in a large empty room, the sound echoes and re-echoes for a considerable time. This re-echoing phenomenom is called reverberation. The non-homogenous particles in the sea which reflect and re-radiate sound are called scatterers.

Reverberation varies from place to place in the ocean. It also varies from time to time in the same locale as a result of layers of scatterers that change depth with temperature variations in the sea or with sea state. And, of course, it varies with changes in the surface boundary of the sea.

Reverberation is a function or results from an active sonar. When the level of reverberation is so great that target echoes are masked in the reverberation, the sonar is said to be "reverberation limited."

Reflection

This is a special case within the general context of scattering. Reflection loss is that decrease in acoustic intensity resulting from change of direction of the wave due to reflection from the surface, bottom or a target, and some absorption of the sound by the bottom.

Ray Path Bending

The rate at which sound travels through water can be expressed as a function of temperature, salinity and pressure (depth). Of these

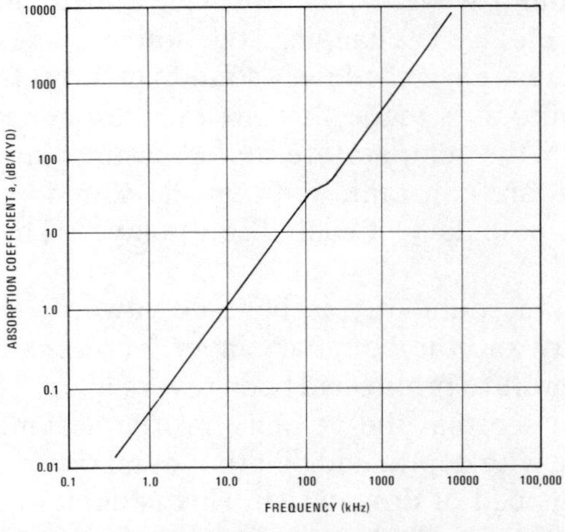

Figure 2-6. Absorption Coefficient in Sea Water of Salinity 35 ppt at 39° F. as a Function of Frequency

factors, temperature is the most variable and most influential. As any of these factors increase, sound velocity increases. As a sound wave travels through a medium, it will bend towards the area which permits the lower sound velocity. To illustrate, a car traveling along a highway that gets its right wheels into a soft shoulder will be pulled to the right. The soft shoulder slows the right wheels. Much the same happens to sound waves in water. The properties of the particular area of the ocean we are operating in and its sound velocity profile (SVP) determine the possible paths which sound takes in traveling from the source to the detector. These paths vary from a simple straight line/shortest distance path to bottom bounce or convergence zone paths. In addition, sound ducts or channels occur frequently in the ocean. Of these paths, some are more reliable then others. As used in this context, reliability refers to these characteristics: no boundary reflection and predictable behavior.

Ray Path Theory

Ray path theory is the tracing of sound waves or rays in the ocean. It is based on the concept of sound waves as previously discussed, and on the bending of sound rays as a function of change in ocean temperature and on ocean depth. It has already been stated that sound velocity increases with temperature and pressure (depth), and that when a sound wave travels through a medium of conditions where sound velocity is changing, the sound wave is refracted or bent towards the area of the lower sound velocity. In tracing sound rays, an assumption is made that the medium consists of discrete layers in which the temperature and pressure, and therefore the sound velocity, are constant, and that the sound velocity changes abruptly at the boundary of each discrete layer. This is depicted in figure 2-7.

Note that as the sound ray is bent downward, the angle made between the ray and the boundary layer increases.

If the temperature profile had been reversed, i.e., increasing with depth, one can see that the velocity profile assumed would have been increasing with depth, and that the sound rays would have been bent upward instead of downward. This is depicted in figure 2-8.

This ray tracing is in fact an application of a law called Snells law, which states a mathematical relationship between the velocity (V) of

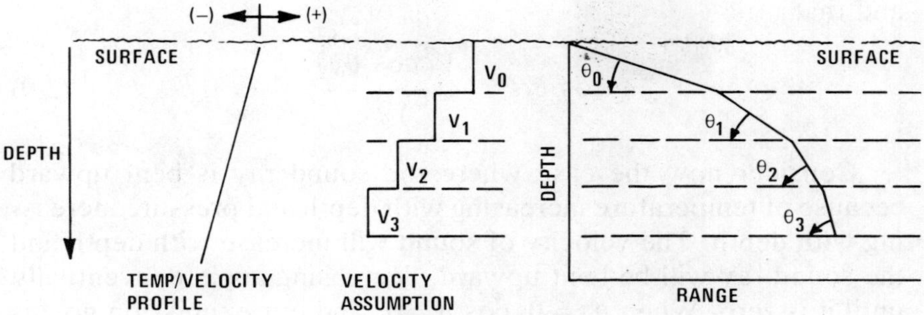

Figure 2-7. Typical Ray Trace, Temperature Decreasing with Depth (Negative Temperature Gradient)

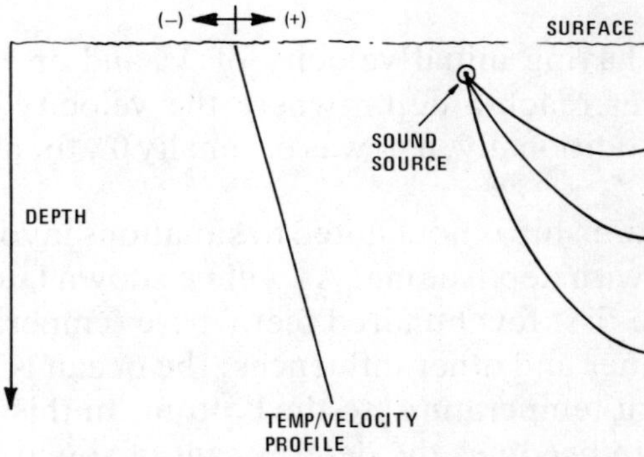

Figure 2-8. Typical Ray Trace, Temperature Increasing with Depth (Positive Temperature Gradient)

sound at the boundary of the layers assumed as described above, and the angle (θ) the sound ray makes with that assumed boundary of the layer. Referring to figure 2-7, Snells law states that

$$\frac{V_0}{\cos\,\theta_0} = \frac{V_1}{\cos\,\theta_1} = \frac{V_2}{\cos\,\theta_2} = \frac{V_3}{\cos\,\theta_3} \tag{2.7}$$

This relationship, is especially useful in cases where temperature increases with depth in computing the depth at which the sound ray is refracted or bent upward. From the foregoing, it can be seen that the velocity

$$V_1 = \frac{V_0 \times \cos\,\theta_1}{\cos\,\theta_0} \tag{2.8}$$

and that

$$\cos\,\theta_1 = \frac{V_1 \cos\,\theta_0}{V_0} \tag{2.9}$$

Consider now the case where the sound ray is bent upward because of temperature increasing with depth and pressure increasing with depth. The velocity of sound will increase with depth and the sound ray will be bent upward, decreasing angle θ_1 eventually until it is zero. When $\theta_1 = 0$, $\cos\,\theta_1 = 1$, and our expression now is written:

$$\cos\,\theta_1 = 1 = \frac{V_1 \cos\,\theta_0}{V_0}, \text{ or } V_1 = \frac{V_0}{\cos\,\theta_0} \tag{2.10}$$

This tells us that a ray having initial velocity of V_0 and an initial direction of θ_0, can never reach a depth where the velocity V_1 is greater than $V_0/\cos\,\theta_0$. Additionally, if θ_0 was originally 0°, then $V_1 = V_2$.

The value of this relationship is not limited to situations involving increase in temperature with depth alone. As will be shown later, in the deep ocean, after the first few hundred feet where temperature varies according to weather and other influences, the ocean is basically isothermal (constant temperature) to the bottom. In this deep isothermal zone, pressure becomes the determinant of sound velocity. Remembering that sound velocity increases with depth, it can be seen that eventually a depth would be reached at which the relationship $V_1 = V_0/\cos\,\theta_0$ would prevail. At this point, the sound path would be bent upward. The sound path thus formed is called a convergence zone (CZ) sound path, and is described more fully in following sections.

In the discussions following, and in viewing ray path tracings in general, it is appropriate to keep in mind that while the traces may only show a few sound rays, for most sound sources such as active sonar projector signals or the radiated machinery noise of a submarine, the sound radiated is multidirectional in both horizontal and vertical directions. Submarine machinery could be expected to radiate more or less equally in all directions, similar to the light rays from an unshaded light bulb. Active sonar projector radiations might be likened to a flashlight beam. In either case, more than one ray is involved.

Principal Sound Paths in the Ocean

There are several principal sound propagation paths of interest to the ASW Tactical Commander. These are described below.

Direct Path

Direct path is, as the name implies, a straight line path direct from source to detector. The direct path is affected by all of the sources of energy loss described above. With a positive temperature gradient (temperature increasing with depth), the direct path sometimes provides some very unusual sound propagation when the sound rays are reflected upward and thus "trapped" in a surface duct. While ducting is frequently observed, the precise tactical value is difficult to predict. This will be discussed in greater detail later.

Bottom Bounce Path

The bottom bounce sound path is a path utilizing sound energy either beamed or bent towards the ocean bottom and results in sound energy being bounced (reflected) off the ocean bottom. While this path of sound energy travel is relatively free from bending due to thermal conditions when beamed, it is highly dependent upon depth and the absorption of sound by the ocean bottom. To illustrate the depth dependency, a beam depressed 15° from the horizontal will bounce a ray off the bottom about 12,000 yds. distance in 1500 fathoms of water, thus providing a detection range of about 24,000 yds. If the depth is reduced to 150 fathoms, the detection range reduces to about 2400 yds. Propagation has been observed with 2 or 3 bounces off the bottom, with corresponding increase in detection range. The value of the sound path is highly dependent upon the reflectivity of the bottom, i.e., how much additional propagation loss is suffered because of absorption by the bottom, and how level the bottom is. The ocean bottom in some areas of the world is much more reflective or more level than in others. The bottom bounce sound path does not meet the criteria for reliability as defined above under "Ray Path Bending."

Convergence Zone Sound Path

Convergence zone sound path, as discussed above, occurs in very deep water, where the effect of pressure results in increased sound velocity until sound rays are bent upward.

For a shallow source and a shallow detector in the deep ocean, the convergence zone distance between them may be as much as 35 miles. In the deep ocean, the sound energy from the source would be refracted or bent downward as a result of decreasing temperature until the increase in pressure bends the sound rays upward. When the sound rays approach the surface at the detector end of the convergence zone, two effects are noted. The first is a focusing effect or gain, which increases the energy seen by the detector. The second is the subsequent downward bending of the focused energy due to negative temperature gradient (decreasing temperature). If the original sound source energy is strong enough, the signal may be detectable at distances of two or three convergence zones, or more.

The direct path, bottom-bounce, and convergence zone sound paths are shown graphically in figure 2-9, along with some nominal values for spherical spreading loss and range. For simplicity in this figure, the bottom-bounce path is depicted as a single beam, whereas in the real world, a wider beam corresponding to a multidirectional sound source would be the normal case.

In figure 2-9, the convergence zone path is shown as having depth based on the concept of a wide sound beam or "bundle" of sound rays emanating from a source. By definition, the convergence zone is formed when the upper ray of the bundle becomes horizontal, i.e., at the depth where the sound velocity is equal to the higher of the velocities at either the surface or of the bottom of the surface layer. As the sound energy travels upward and approaches the surface, the path narrows, tending to focus the sound energy. In the annulus of the convergence zone as depicted in figure 2-9, focusing or convergence gains of from 10 to 25dB have been reported.[1]

The difference in depth between the ocean bottom and the depth

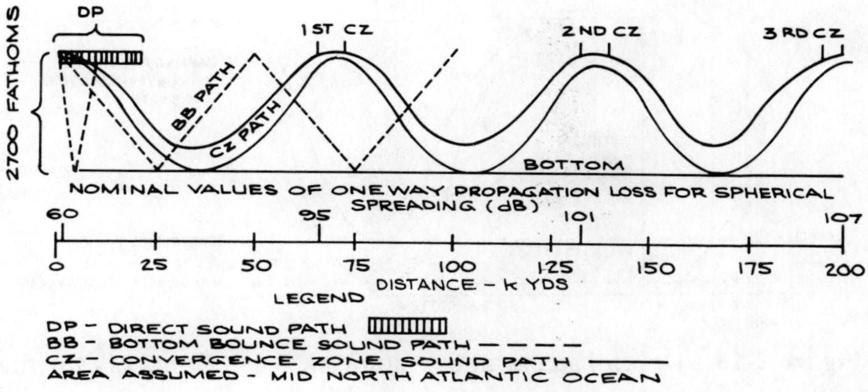

Figure 2-9. Three Typical Sound Paths Between a Source and a Receiver in Deep Water in Mid-North Atlantic Ocean and Nominal Values of One-Way Propogation Loss

at which the upper ray becomes horizontal is called depth excess. It can be seen that the depth excess available defines the depth excess of the bundle of rays or "amount of sound" that forms the convergence zone sound path. Depth excess is important to the reliability of the convergence gains noted above. According to the New London Laboratory, Naval Undersea Systems Center, New London, Connecticut, a depth of 1200 ft. is required to form an operationally useful convergence zone, and 1800 ft. depth excess is preferred.

Remembering that velocity increases with depth, the corresponding amount by which the velocity of the bottom ray exceeds that of the top ray in the convergence zone path bundle is called velocity excess. These concepts are depicted in figure 2-10.

Sound Channel Propagation Paths

The sound velocity structure of the sea, combined with its surface and bottom boundaries causes sound channels or ducts to occur in the sea. In general, a channel or duct will occur whenever sound traveling horizontally from a source is prevented from spreading vertically and is confined within the boundaries of the channel. Some channels are formed on the basis of velocity structure alone, i.e., refraction, while others result from a combination of refraction

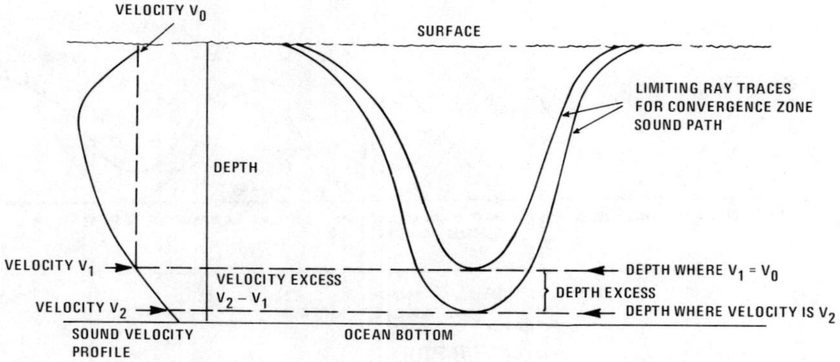

Figure 2-10. Depth Excess and Velocity Excess Concepts for the Convergence Zone Sound Path

due to velocity structure, and reflection from the surface or bottom boundaries of the ocean. A discussion of such channels is more meaningful against a background of a typical velocity structure of the sea, which is reviewed in the next section.

The Deep Sea Velocity Profile

Figure 2-11 is a typical velocity profile of the deep ocean. As can be seen, this profile is nothing more than a plot of sound velocity vs. ocean depth. The profile is usually divided into layers, each having specific characteristics that differentiate it from the others. As might be expected, the sound transmission characteristics of each layer are different. In figure 2-11, some of these layers are labeled "thermocline"—a word meaning a layer of the ocean in which temperature changes with depth.

Surface Layer

Proceeding from the surface downward, the surface layer is heated daily by the sun and mixed by the wind. Mixing by the wind is frequently so thorough that this layer is of constant temperature (isothermal) from the ocean surface to the bottom of the layer. The depth of the mixed layer is directly proportional to the magnitude of the wind driven wave action; the more violent the frontal or storm passage, the deeper the layer. The writer has measured isothermal

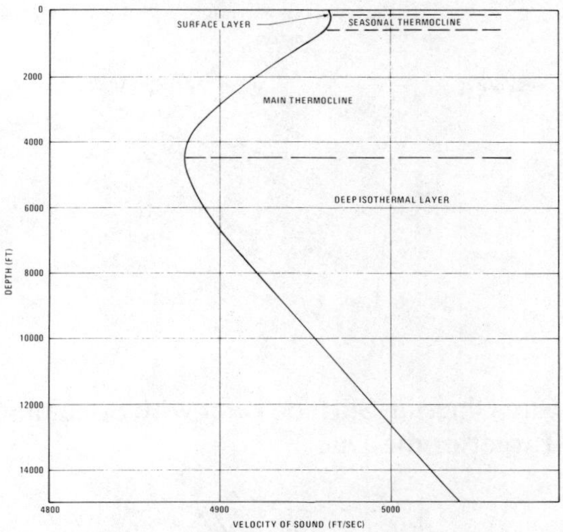

Figure 2-11. Typical Deep Sea Velocity Profile and its Principal Layers

layers to depths of 300-600 ft. in the Mediterranean and in the Atlantic.

Sound emanating from a source in an isothermal layer is trapped in the layer and "ducted" so to speak. When the submarine target and the searching sonar are both in the mixed layer, exceptionally good sound propagation is observed, and target detections at longer range than normal are the rule.

When the mixed surface layer is so shallow that only the searching active sonar is located in it while the submarine target is below the layer, a marked reduction in detection range is experienced due to some of the sound energy being trapped in the layer, and other sound energy being refracted downward near the source, thus creating a below layer shadow zone. Ray path diagrams for the conditions described above are shown in figures 2-12 and 2-13.

In tropical areas in particular, the upper part of the mixed surface layer will be heated by the sun. The increasing heat of the sun as the day wears on causes a shallow positive thermal gradient in the surface layer. Also, as the wind comes up in force as the day wears on, the top of the surface layer becomes mixed and is isothermal. The net effect is that by mid-afternoon, a positive velocity gradient exists in the surface layer, trapping the emissions from an active sonar in the shallow surface duct.

Other local transient phenomena influence sound propagation in

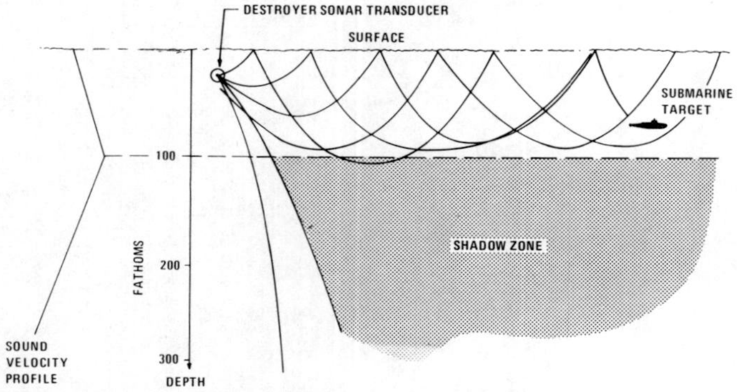

Figure 2-12. Ray Diagram Surface Duct with Sonar and Submarine Target in the Duct

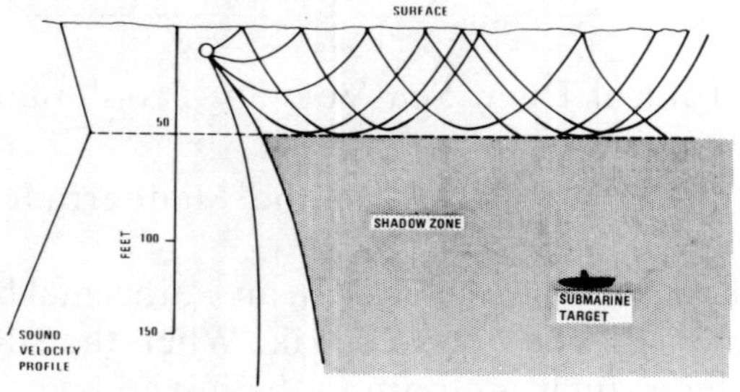

Figure 2-13. Ray Diagram with Sonar in the Surface Duct and the Target Below the Shallow Surface Layer

the sea. Rain contributes to background noise. While this tends to reduce own sonar performance, it can also decrease detectability of own force radiated noise signature.

Seasonal Thermocline

Just below the surface layer is the seasonal thermocline. In this layer, the temperature decreases with depth. The depth at which the decrease begins is dependent upon the season of the year. For example, in the stormy winter months in the North Atlantic Ocean or the Mediterranean Sea, the surface waters are churned and mixed by the wind to greater depths than in the summer. During these winter months, the bottom of the mixed layer may become so deep as to cause the seasonal thermocline to lose its identity. The superior sound transmission qualities of the deep mixed layer have already been discussed.

Main Thermocline

The next deeper layer in the sea is the main thermocline. This layer is characterized by a reduction in temperature with depth and is little affected by variations in surface conditions. The bottom boundary of this layer is the top of the deep isothermal layer, or the depth at which the temperature of the sea is nearly constant at 39°F. In the deep isothermal layer, the velocity of sound increases with depth as a function of increased pressure.

Major Ocean Currents

A final environmental factor of which the ASW commander should be aware is the abrupt change in sound velocity and the sound propagation that may be experienced when operating on the boundaries of such pronounced currents as the Gulf Stream, the Japanese Current, and the Labrador Current. The boundaries of these currents are not always sharp and well defined but rather tend to form delta-like fingers, with attendant abrupt changes in layer depth, velocity structure, and sound propagation.

The Deep Sound Channel

The deep sound channel is formed in the ocean as a consequence of the sound velocity structure of the ocean. This is illustrated in figure 2-14. All rays leaving the projector between the two limiting rays A and B are alternately refracted up and down between the indicated boundary lines according to Snells law previously illustrated. This sound channel provides excellent propagation of sound. It is also called the SOFAR channel (SOFAR means *so*und *f*ixing *a*nd *r*anging), used in the rescue of downed aviators. The downed aviator

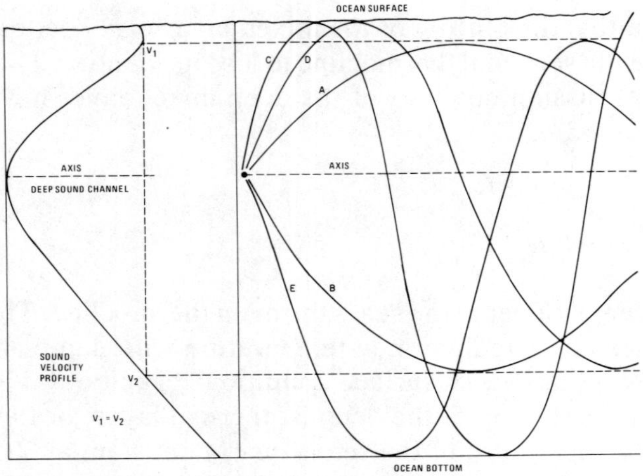

Figure 2-14. Sound Velocity Profile and Typical Ray Path Diagram for a Sound Source (or Receiver) on the Axis of the Deep Sound Channel

released a small explosive charge set to detonate at or near the axis of the deep sound channel. Capable of being detected for thousands of miles at listening shore stations, the downed aircraft could be located based on time of arrival of the explosive signal at two or more stations. Figure 2-14 also shows the paths that might be followed for sound rays not bounded by rays A and B. These may follow propagation paths that are reflected at the sea surface and bottom (ray C), or refracted and reflected from either the sea bottom or the sea surface (rays D and E). Ray D is commonly called the refracted surface reflected (RSR) path, while ray E is called the refracted bottom reflected (RBR) path.

For sound transmission in a channel or duct, a modification to the spherical spreading loss expression (equation 2.5) is required to express propagation loss quantitatively. For a situation where the depth of the duct is small compared to the range over which the sound is transmitted, the analogy of flow through a cylindrical pipe is appropriate. The propagation loss for cylindrical spreading is given by the expression:

$$\text{Cylindrical Spreading Loss (dB)} = 10 \log R. \qquad (2.11)$$

where R is the range in yards.

Comparison of expression (2.11) with that for spherical spreading (equation 2.5) indicates that the propagation loss experienced under ducting conditions may be only one half as great as the losses normally expected due to spherical spreading.

If all of the potential sound rays emanating from a source, or located on the axis of or within the deep sound channel are visualized, it is apparent that some of these rays will intersect and combine with each other at a hydrophone located within the channel. These regions of intersection and partial focusing are called caustics. Under the special conditions where a caustic occurs at or near the sea surface, the area of focusing gains is called a convergence zone.

The upper and lower boundaries of the deep sound channel are not fixed in depth, but are determined only by the sound velocity profile and the ocean depth. Thus, the channel may in fact extend from the sea surface to the sea bottom at some times in some areas.

The Reliable Acoustic Path

Review again the concept of the formation of the convergence zone sound path for a shallow source of receiver. The sound energy is refracted downward due to decrease in temperature until the deep isothermal layer is reached, at which time pressure due to depth increases the velocity of sound. When the sound velocity below the axis of the deep sound channel equals or exceeds the maximum near surface sound velocity, the sound energy path is refracted upward to or near surface.

If the source or receiver in the above case were placed at or below the depth at which the sound energy is refracted upward, sound energy from the projector or to the receiver would follow a sound path similar to the previously described convergence zone sound path. This path is called the reliable acoustic path (RAP). It is so termed because it is not dependent on bottom or surface reflection losses, nor on the variable conditions previously described as occurring in the near-surface regions of the sea. Figure 2-15 is a conceptual RAP ray diagram based on a typical North Atlantic sound velocity profile.[2]

Shallow Water Sound Propagation

Shallow water generally refers to water depths of less than 1000

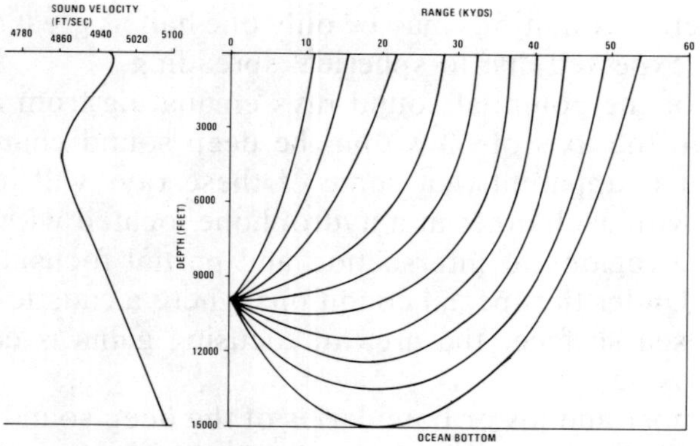

Source: *A Reliable Acoustic Path Handbook,* U.S. Naval Air Systems Command (Washington, D.C.: U.S. Navy Department, 1971).

Figure 2-15. Sound Velocity Profile and Corresponding Ray Path Plot for a Sound Source (or Receiver) at 10,000 Ft. Depth

fathoms. Propagation of sound is very difficult to predict in shallow water because of multiple reflections of sound from bottom to surface to bottom. In a sense, shallow water is like a duct, but the reflection losses are not precisely predictable. As indicated above, the principal mode of sound propagation is via multiple bottom and surface reflection. In this situation, it is possible that active sonars could achieve longer than normal detection ranges even under negative gradient conditions. Further, target depth has little influence on sonar detection performance in shallow water areas.

Measuring the Tactical Sonar Environment

The principal assets available to the ASW commander to measure the tactical sonar environment are data from past marine geophysical surveys, past measurements made by other ships in the same area at the same season, and the bathythermograph or velocimeter furnished all ASW ships today. Data taken in the past, correlated with the current weather and current season, can present a general picture of what to expect in the way of velocity profile, or at least of the boundaries to be expected during a given period. These data are of value to the staff ASW Operations Officer in the planning stages of the operation. After the force or units are at sea, the bathythermograph or velocimeter provided is the basis for establishing the actual velocity profile of the ocean. Based on the variability in the sea with time, discussed above, it is obvious that a policy of regular periodic measurement of the velocity profile is required during the course of the ASW operation. Normally, one or two ships in the force are assigned the bathythermograph guard duty. These ships periodically measure the temperature or velocity profile, and promulgate it to all ASW units in company.

The bathythermograph (BT) is a temperature recording device dropped or lowered into the water, in which is recorded temperature vs. depth. The velocimeter is an instrument similarly deployed, which records directly, sound velocity vs. depth.

In addition to these measurements, it is possible to measure propagation loss directly. Given two ships at sea, one with a calibrated sound source and the other with a signal level meter, it is feasible to measure the actual propagation loss in the sound path between these ships. If the sound source has sufficient output power, one could expect to measure the bottom bounce and convergence zone sound paths as well as the direct path.

The manner in which temperature or sound velocity profile is combined with sonar design performance to predict expected tactical sonar performance will be discussed in Chapter 3.

Notes

1. R. J. Urick, *Principles of Underwater Sound for Engineers* (New York, N.Y.: McGraw-Hill Book Co., 1967), p. 137.

2. *A Reliable Acoustic Path Handbook,* U.S. Naval Air Systems Command (Washington, D.C.: U.S. Navy Department, 1971).

3 The Sonar Equations

In Chapter 1, it was stated that Tactical Sonar Performance was a function both of sonar equipment performance and the sound propagation characteristics of the ocean medium. The basis for this statement is;

a. that sonar equipment is designed to detect the arrival of a certain minimum sound energy level at the sonar, and
b. that the distance or range over which the sound energy travels to produce this minimum detectable signal (MDS) is dependent upon how much energy is lost en route from the sound source to the sonar.

It was further indicated that a relationship called the sonar equation was available to combine these two factors, sonar design and ocean propagation loss, in producing a quantitative measure of tactical sonar performance in kilometers or kiloyards of range.

In Chapter 2, those characteristics of the ocean environment which impede, divert, or diminish sound energy over the several sound paths of interest to the ASW Tactical Commander were discussed, along with the variability in some of these factors from time-to-time and place-to-place (the temporal and spatial variation of sound propagation in the ocean.)

This chapter will deal with those design characteristics of sonars which are used as a quantitative measure of sonar design and how this quantitative measure of sonar design performance is combined with quantitative measures of ocean propagation loss to produce a quantitative statement (in units of range) of Tactical Sonar Performance.

Before proceeding directly with the sonar equation, there are three additional thoughts to establish in mind. These are: the frequency dependence of sound intensity levels, spectrum levels, and relative sound intensity levels.

Frequency Dependence of Sound Pressure Levels.

In Chapter 2, it was indicated that the principal losses of acoustic signal energy in the ocean medium were due to spherical spreading, and absorption or attenuation. Based on this, one could state that:

$$\text{Propagation loss} = \text{Spreading loss} + \text{absorption loss.} \quad (3.1)$$

Considering again the stone dropped in the pool, observations indicate that as the wave travels farther from the source, it eventually loses its identity and blends in with the previously existing wave motion of the pool. This indicates that the spherical spreading loss is essentially a function of distance or range from the source, and is not a function of any other attribute of the wave such as frequency or wavelength. In the decibel notation, the loss of acoustic signal level as a function of distance or range traveled is simply expressed as:

For spherical spreading.

$$\text{Spreading Loss (dB)} = 20 \log R, \quad (3.2)$$

For cylindrical spreading.

$$\text{Spreading Loss (dB)} = 10 \log R \quad (3.3)$$

where R is the range or distance in yards. This relationship for spherical spreading is depicted graphically in figure 3-1.

However, from Chapter 2, we know that absorption loss does vary with the frequency of the sound energy being transmitted, and the curve in figure 2-6 indicates the value of the rate at which absorption loss varies. In acoustic calculations, the absorption loss coefficient is symbolized by the letter a and is expressed in decibels per kiloyard. Absorption loss is calculated as follows:

$$\text{Absorption loss (dB)} = aR \times 10^{-3} \quad (3.4)$$

where R is range or distance in yards.

Propagation loss (N_W) may now be expressed as follows:

$$N_W \text{ (dB)} = 20 \log R + aR \times 10^{-3} + A \quad (3.5)$$

where A is the transmission anomaly factor, which accounts for abnormalities such as irregular refraction and reflection of sound within the ocean medium. If one considers that the sum of spreading loss and absorption loss to be the predicted transmission loss, then A is the difference between predicted and measured transmission loss.

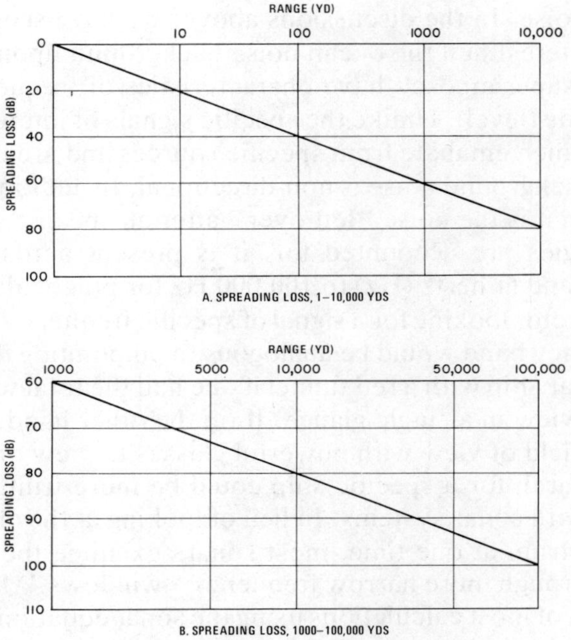

Figure 3-1. Spreading Loss as a Function of Range

Figure 3-2 shows the effect of absorption loss when added to spherical spreading loss.

The significance of the frequency dependence of propagation loss is that when calculating the acoustic signal resulting from a signal of interest superimposed on the ambient background noise of the ocean, or the ratio of the signal level to noise level, this calculation must be made using signals of the same frequency to be valid (i.e., we add apples and apples, etc.). Further, since sonar systems are designed to operate in specific frequency bands, calculations for these systems must relate to the design frequency band of the specific sonar to be valid.

Sound Spectrum Levels

One of the problems of a sonar or sound detector is in discriminating between acoustic energy signals of interest and background or ambient area noise. In the discussions above, we have seen that both signals of interest and the ocean noise background upon which the signals are superimposed have characteristics of frequency (tone) and amplitude (level). Unlike the specific signals of interest in sonar and ASW which emanate from specific sources and are directional, the ocean background noise is non-directional. In addition, because by definition it is the noise "left over" after all specific identifiable sound energies are accounted for, it is present across the total frequency band (1 hertz (Hz) to 100,000 Hz for practical purposes). A sonar detector looking for a signal of specific frequency across this total frequency band would be analogous to pinpointing the location of a particular ship with a red funnel if one had the total world ocean shipping in view in a single glance. If on the other hand, one could narrow his field of view with powerful glasses to view one ship at a time, the search for a specific ship could be more fruitful.

So it is with sonar systems. In lieu of looking at the entire sound energy spectrum at one time, most sonars examine the frequency spectrum through more narrow frequency "windows." In addition, for purposes of most calculations using the sonar equation, signals of interest and background sound levels are expressed in terms of amplitude or sound pressure levels in a frequency band 1 Hz wide. Values thus expressed are termed *spectrum levels*. Figure 3-3 portrays graphically the spectrum levels for the deep ocean ambient noise. Also illustrated in this figure are the increases in noise level resulting from ocean shipping and the increased molecular activity associated with increased wind-generated wave action.

The Signal-to-Noise Ratio (S/N).

The concept of signal-to-noise ratio (S/N) is fundamental to the sonar equations. Consider a situation where own ship has a simple passive sonar as shown in the upper segment of figure 1-2, and a target submarine is located some distance from own ship, closing in. The mission of our sonar operator is to detect the radiated noise signal of the target submarine. Remembering that there is a background noise level in the sea at our own hydrophones and assuming the target initially is at such a range that the noise level is greater than the signal level arriving at own hydrophones, the operator hears only the background noise. As the target closes, we know from equation

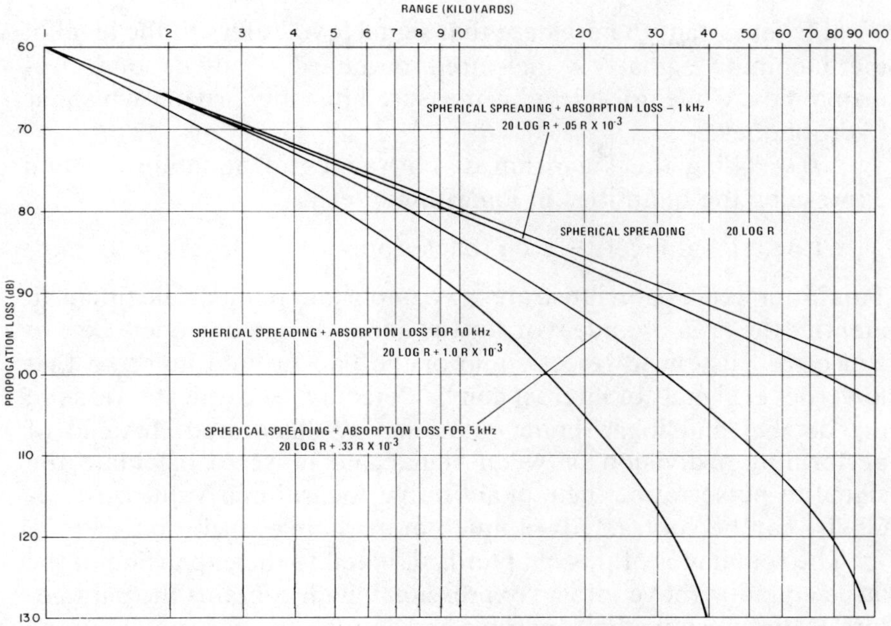

Figure 3-2. Plot of Range vs. Propagation Loss, Showing the Magnitude of the Difference Between Spherical Spreading Loss Alone, and Spherical Spreading Loss Plus Absorption Loss

3.5 and figure 3-1 that the target signal level will slowly increase. When the target has closed to a situation where the average target signal level just equals the average level of the background noise, the operator will have an opportunity to detect the target signal according to some preestablished success criteria specified in terms of probability of detection desired, and rate of false alarms deemed acceptable (more about this later). Stated in other terms, a detection occurs when

$$\text{Signal level} = \text{Background noise level}. \tag{3.6}$$

or expressed as a ratio,

$$\frac{\text{signal level}}{\text{background noise level}} \left(\frac{S}{N}\right) = 1. \tag{3.7}$$

It is important to note here that signal level refers to the level of the incoming signal as measured outboard of own sonar hydrophones, while noise level is measured just inboard of own sonar hydrophones.

Now taking the logarithm of both sides of equation 3.7 and expressing the quantities in logarithmic terms,

$$\text{Log (signal level)} - \text{Log (background noise level)} = 0. \quad (3.8)$$

This latter expression illustrates two important points. The first is to identify the area of endeavor for the sonar design engineer, i.e., to cause the system to reject some of the background noise so that lower level signals of interest can be detected. Second, the value of the decibel and logarithmic notations are illustrated. Instead of performing a division between signal and noise to calculate the signal-to-noise ratio, their quantitative logarithmic values, in decibels, can be subtracted—a much more simple operation.

The remainder of this chapter is devoted to the expansion of the basic equality above into an expression which contains the parameters of the sonar systems (passive and active), the parameters of the sea, and the parameters of the target.

The Passive Sonar Equation

Signal Level (L_S)

Consider first the target radiated signal level. This signal is a composite of a number of different individual sounds generated by shipboard main propulsion and auxiliary machinery or by hydrodynamic sounds such as propeller cavitation and water flow over the hull. These different sound sources result in a target radiated noise signature composed of both discrete frequencies and broad band noise. This target-generated parameter is conventionally labeled L_S and is expressed quantitatively in decibels as though it had been measured at a distance of one yard from the target (considered as a point source).

Propagation Loss (N_W)

In traversing the distance between the radiating target and the

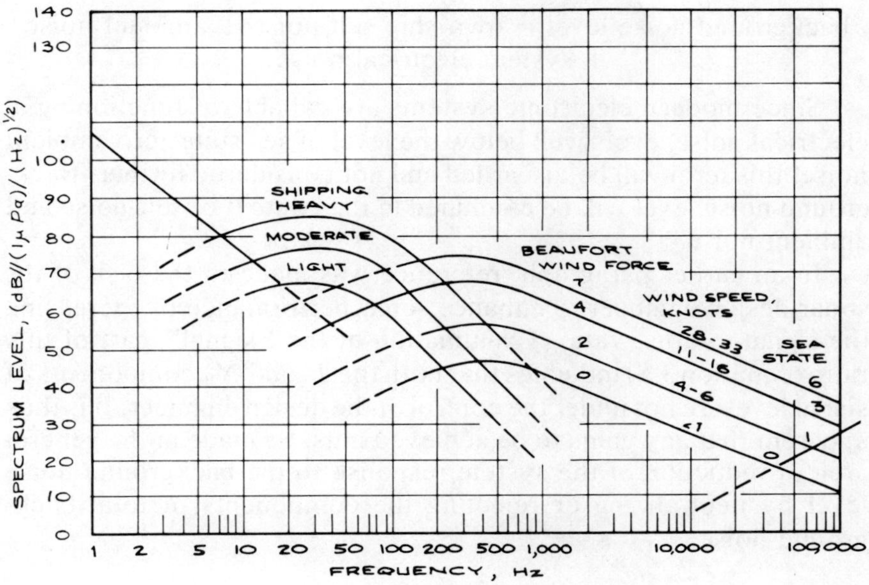

Source: R.J. Urich, *Principles of Underwater Sound for Engineers* (New York: McGraw-Hill Book Company, 1967), figure 7.5, p. 168.

Figure 3-3. Average Deep-Water Ambient-Noise Spectra (LN Component)

searching sonar hydrophones, the radiated signal level, L_S, is reduced in intensity by the propagation or transmission loss of the sea (equation 3.5). Propagation loss is labeled N_W and is expressed in dB. It is thus apparent that the signal level arriving at the sonar hydrophones can be stated as $L_S - N_W$. This completes the "signal" part of the basic equality stated above. Substituting in equation 3.8, we now write:

$$L_S - N_W - (\text{background noise level}) = 0. \qquad (3.9)$$

Background Noise Level (L_N)

The background noise level against which the sonar system must prevail if the system is to detect target signals may be considered to consist of three principal components, viz, background noise level = own ship self-noise + ambient noise + system electrical noise.

Since modern electronic systems are capable of functioning at electrical noise levels well below the level of sea state zero ambient noise, this term will be discarded and not considered further. Background noise level will be examined in the context of self-noise and ambient noise.

In an earlier paragraph, reference was made to the task of the sonar design engineer to enhance sonar performance by increasing the signal-to-noise ratio. Examination of the "signal" part of this ratio (equation 3.8) indicates that both the L_S and N_W components of signal level are not under the control of the design engineer. It is thus apparent that any gains to be achieved must be made in the general area of reduction of the system response to the background noise level by neutralizing or rejecting the components of total background noise.

Directivity Index (N_{DI}). Consider first the ambient noise background of the sea itself. This noise arises principally from the wave motion of the sea, from marine creatures, and from the radiated noise of ocean shipping. This component of background noise is isotropic (non-directional). Consider now a single omnidirectional hydrophone seeking a directional signal from a target in an isotropic noise field. The hydrophone sees noise over an arc of 360 degrees, while the signal sought is available on only a single direction (line of bearing). It is apparent that if one could arrange a number of hydrophones and "steer" them to look in a single direction, noise in all other directions would be rejected and the signal-to-noise ratio in any selected direction would be enhanced. The sonar design engineer does this in the manner illustrated in figure 3-4.

In figure 3-4, let line $E_0 - E_x$ represent a line of array elements $E_1 - E_n$ spaced d distance apart. Let lines a $a - a^1$ through $f - f^1$ represent the successive positions of a plane wave of sound energy traveling at velocity c, as it strikes the hydrophone elements of array $E_0 - E_x$ from a direction θ with respect to the array axis. Let L be the distance the sound wave travels from the time it strikes one array element until the next. Consider that as the sound wave strikes each element, it generates a voltage proportional to the intensity of the incoming signal. It can be seen that the output voltage of element E_2

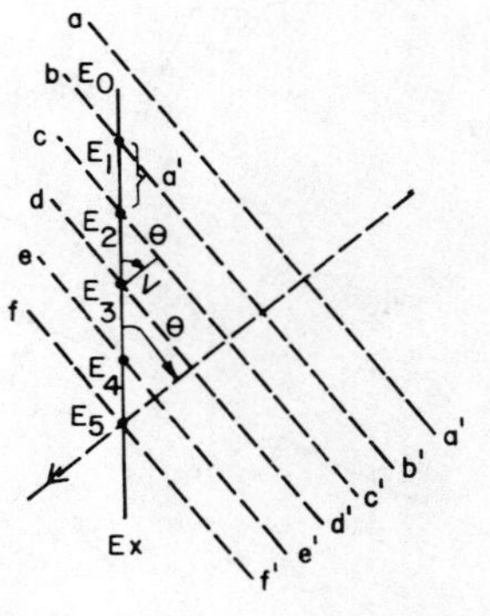

Figure 3-4. Directional Beamforming with a Simple Line Array

relative to E_1 will be delayed at the processor by the length of time required for the sound to travel the distance L which will be equal to $d \sin \theta$, and that for successive elements, the delay for the nth element with respect to element $E_1 = nd \sin \theta$, where n is the number of the hydrophone in the array. By electrically delaying the arrival of the signal from each element to the processor according to its position in the array, and then adding these electrical signals, it can be seen that the signal-to-noise ratio for the array input to the processor will be greater than that input by a single hydrophone. This is true because the sinusoidal sound signals added or summed will always be in phase at all the array elements, while the background noise is out of phase. At the same time, the directionality of the array reduces noise arriving from other directions relative to the signal. These considerations are the basis for computation of the parameter directivity index (N_{DI}) in the sonar equation. For the simple line array illustrated, N_{DI} is calculated:

$$N_{DI} = 10 \log 2L/\lambda \qquad (3.10)$$

where L = array length, and

λ = the wave length for the frequency concerned

For most hull mounted sonars, the active and receiving hydrophone arrays are arranged on the surface of a cylinder, although some submarine sonar arrays are installed on the surface of a sphere. Directionality, in these cases, is established by reducing the arc of the cylinder or sphere to an equivalent virtual line array by a simple

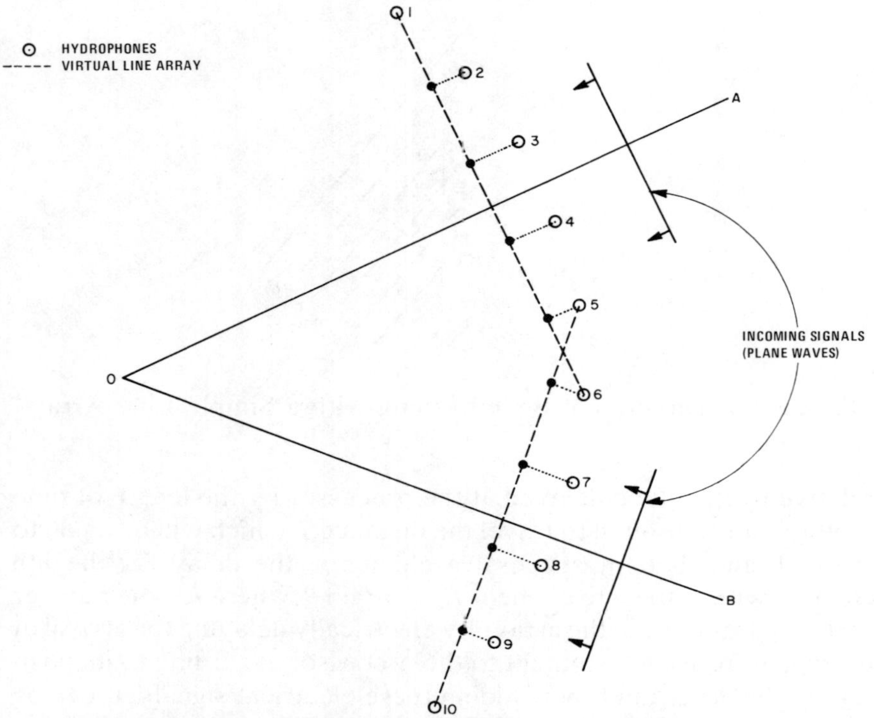

Figure 3-5. Direction Beamforming with a Curved Array (Arc of a Cylinder or Sphere)

phase delay between the hydrophones and then summing the hydrophone outputs. Figure 3-5 illustrates this process.

Shown in figure 3-5 is a 10 hydrophone array arranged in an arc from the surface of a cylinder or sphere. Incoming plane wave signals are shown approaching the array from directions AO and BO, respectively. For direction AO, a virtual line array normal to direction AO is formed using elements 1-6 inclusive. A similar virtual line array is formed using elements 5-10 inclusive for the signal from direction BO. In summing the signal from the six hydrophones used for each beam, the output from those hydrophones which "see" the signal first is delayed electrically by the time it would take the sound wave to travel the distance from that hydrophone to the line representing the virtual array. The functional task of determining the direction of an incoming signal by forming virtual arrays and selecting the right one is the task of the system beamformer.

By convention, total background noise level is labeled L_N. For this discussion, label the ambient sea noise component L_{NA}. If we subtract from L_{NA} the gain achieved by rejecting part of L_{NA}, i.e., N_{DI}, then this part of the total background noise can be expressed $L_{NA}-N_{DI}$.

Consider now self-noise. Self-noise is caused by the flow of water over the hydrophones and specific machinery noise. It is apparent that this component of background noise is both directional and incoherent (non-sinusoidal) in nature. The sonar engineer neutralizes noise of this nature, again by using an array. In this instance, the output of the several hydrophones are combined in a manner such that the signal-to-noise ratio of the array exceeds that of a single hydrophone. This gain is called Array Gain (L_{SNR}), and for a simple line array is calculated as follows:

$$L_{SNR} = 10 \log n \tag{3.11}$$

where n is the number of elements in the array.

If we label the self-noise component of background noise as L_{NS}, then this component of total background noise can be expressed $L_{NS} - L_{SNR}$. Total background noise (L_N) then can be expressed:

$$L_N = (L_{NA} - N_{DI}) \oplus (L_{NS} - L_{SNR})^{a} \tag{3.12}$$

This is much less complicated than it sounds, as illustrated below. Assume an 18 element array at a speed of 10 kts, where:

$$L_{NS} = 45 \text{ dB//(1 uPa)/(Hz)}^{\frac{1}{2}},$$

$$L_{SNR} = 12.5 \text{ dB (10 log 18)},$$

$$L_{NA} = 68 \text{ dB//(1 uPa)/(Hz)}^{\frac{1}{2}}, \text{ and}$$

$$N_{DI} = 9.5 \text{ dB}.$$

$$L_N = (L_{NA} - N_{DI}) \oplus (L_{NS} - L_{SNR})$$

$$L_N = (68-9.5) \oplus (46-12.5) = 58.5 \oplus 33.5 \text{ dB//(1 uPa)/(Hz)}^{\frac{1}{2}}.$$

Figure 3-6 is a curve which is reasonably accurate for performing power summation calculations. The curve is entered with the difference between the quantities to be combined, and the differential

[a] $\oplus$ is a symbol used to indicate power summation or random phase addition. In this summation process, the total or composite background (L_N) is the logarithm of the sum of the anti-logs of the flow noise and ambient noise components.

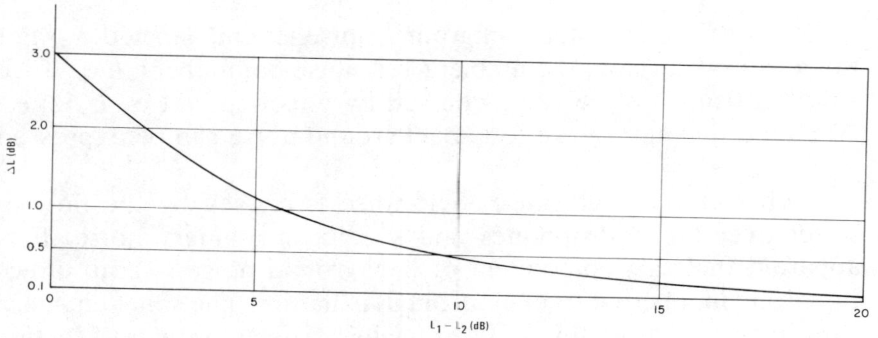

Figure 3-6. Power Summation Curve. L_T (the sum of two power quantities $L_1 + L_2$) $= L_1 + \Delta L$, where $L_1 \geqq L_2$.

quantity (ΔL) to be added to the larger of the two basic quantities is obtained. In this case, the curve is entered with 58.5−33.5, or 25. It can be seen that the larger quantity 58.5, based on sea state, so dominates the situation that the self-noise contributes less then 0.1 dB to the total noise field. By inspection of figure 3-6, one can also see that when both the ambient noise component and the self-noise components are equal, their power sum is increased by 3 dB, which represents a doubling in noise power levels.

For low and moderate speed sonar operation, the ambient noise component generally so dominates that for operational and analytical purposes ambient noise is normally used for background noise and the noise term in the sonar equation is written $L_N–N_{DI}$. The developing sonar equation is thus written:

$$(L_S - N_W) - (L_N - N_{DI}) = 0 \qquad (3.13)$$

$$\text{Signal} - \text{Noise} = 0$$

This form will be used throughout the remainder of this text.

Recognition Differential(N_{RD})

There remains to be discussed one more sonar parameter to make the passive sonar equation complete. This parameter, called Recognition Differential, is labeled N_{RD} and is expressed quantitatively in decibels. Its value is based on the method of presenting the signal-to-noise ratio to the operator (sonar design), the acuity and training of the sonar operator (individual operator capability), and preselected success criteria (user defined). One can see from this that N_{RD} is not only a very important parameter because it provides for expressing quantitatively the man/machine interface, but also is one of the most complex, involving preselected success criteria, sonar hardware design, and human capabilities/variabilities.

Values for N_{RD} nominally can be expected to range from +10 dB to −20 dB where +10 dB applies to the least best trained operator or least best system, and −20 dB applies to the best trained operator or best system.

By arbitrary convention, target recognition or detection is said to occur when 50% of the signals presented to the operator are identified correctly against the prevailing noise background. From this, we can state that Recognition Differential (N_{RD}) is the difference between signal available and background noise when there is a 50% probability of detection of the signal.

The passive sonar equation can now be written:

$$(L_S - N_W) - (L_N - N_{DI}) = N_{RD} \qquad (3.14)$$

Signal − Noise = Recognition Differential

Examining this expression in the context of probabilities that differ from the 50% level, it is apparent that to operate at a higher probability level would require a higher signal-to-noise ratio. While in some cases this could be achieved by placing the best operator in the system and improving N_{RD}, it could also be accomplished by reducing the range between the target and own ship over which the signal must travel, thus reducing N_W. Or, one also might reduce the self-noise component of background noise by reducing speed if own ship noise is greater than ambient noise. All of these tend to increase the signal-to-noise ratio available. In a similar manner, operating at a reduced detection probability permits operating with a less trained operator, or at greater ranges, or at higher speeds.

Figure 3-7 indicates an order of the corresponding change required in signal-to-noise ratio (Signal Excess) as a function of change in recognition probability. It can be seen that to achieve 90% detection probability requires 4 or 5 dB greater S/N ratio than for a 50% probability. Also, if the system is operating 4 or 5 dB down from design, the probability of detection is only about 10%.

There are several other concepts important to the overall context

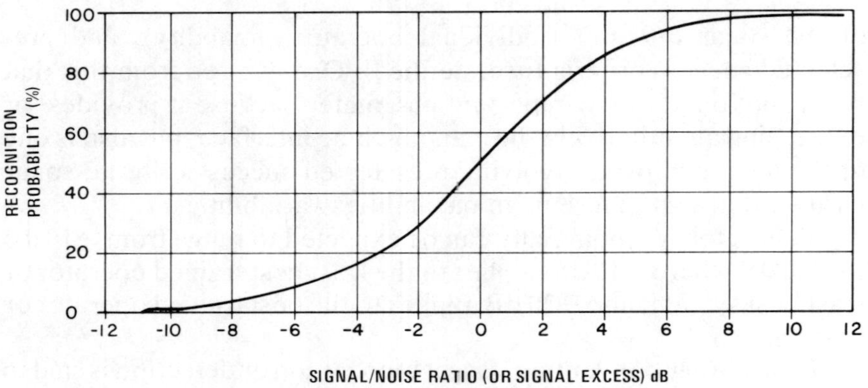

Source: National Defense Research Committee (NDRC), *Principals and Applications of Underwater Sound* (Washington, D.C.: U.S. Government Printing Office, 1946), fig. 10, p. 189.

Figure 3-7. Plot of Recognition Probability vs. Signal-to-Noise Ratio (or Signal Excess)

of the parameter recognition differential. These are: Probability of False Alarm (P_{FA}), Detectability Index (d), Receiver-Operating Characteristic (*ROC*) Curves, and Threshold (T). The probability of false alarm (P_{FA}) may be defined as the probability that a target is reported present, given that a target is not present. Thus success criteria, when related to the required probability of detection (P_D), is important to sonar design.

Detectability Index (d) is a function of the signal-to-noise ratio. Threshold (T) refers to the difference of the signal level above the average noise level necessary for the system or operator to make a correct detection decision. For given values of d (fixed signal-to-noise ratio), different thresholds represent different combinations of probability of detection (P_D) and probability of false alarm (P_{FA}). The relationship between P_D, P_{FA}, and d are summarized graphically on a Receiver Operating Characteristic (*ROC*) curve as shown in figure 3-8. ROC curves, as they are commonly called, are a family of non-intersecting curves, each with a constant value of d plotted against P_D and P_{FA}. Theoretically, a given operator or set of operators, should produce a consistent d for any one value of S/N. Changes in operators, or of operator motivation, or in operator fatigue may cause a shift in threshold along the appropriate d curve, thus changing the values of P_D and P_{FA} at which the system is actually operating.

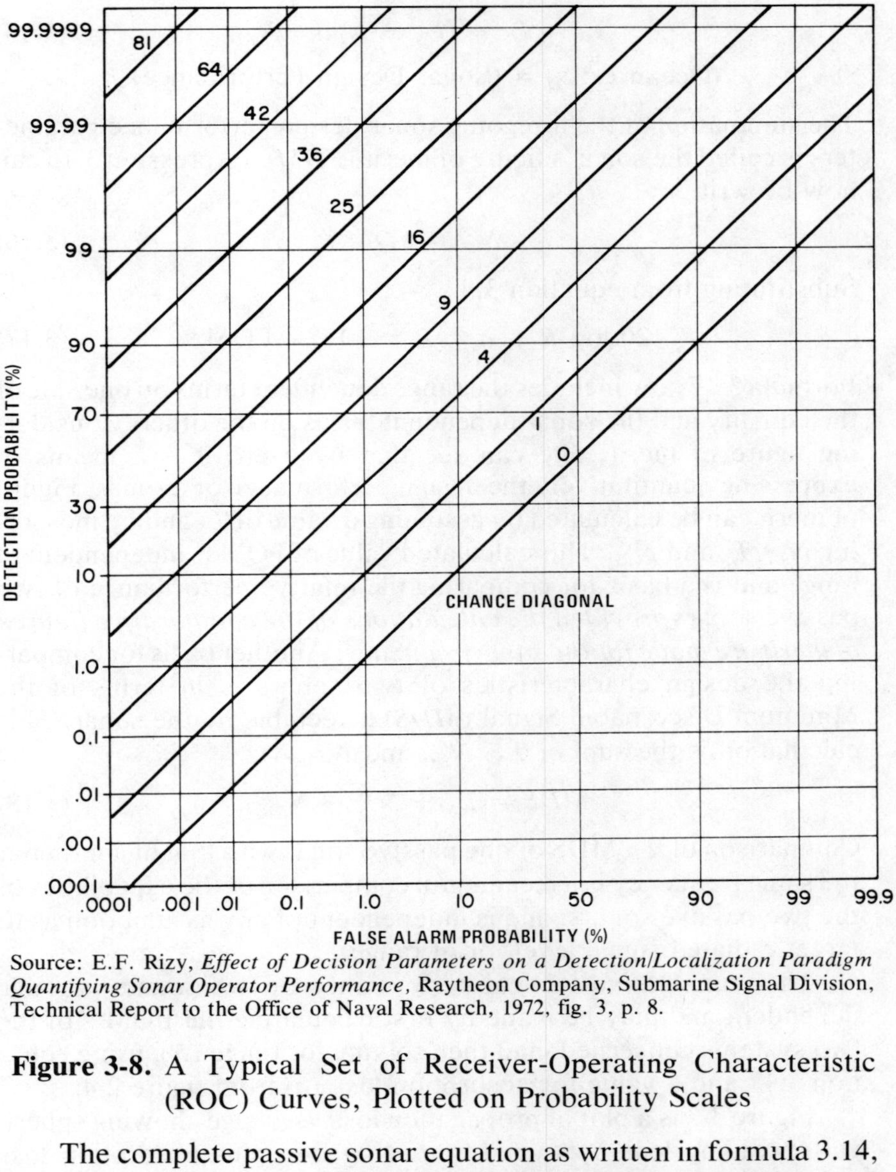

Source: E.F. Rizy, *Effect of Decision Parameters on a Detection/Localization Paradigm Quantifying Sonar Operator Performance,* Raytheon Company, Submarine Signal Division, Technical Report to the Office of Naval Research, 1972, fig. 3, p. 8.

Figure 3-8. A Typical Set of Receiver-Operating Characteristic (ROC) Curves, Plotted on Probability Scales

The complete passive sonar equation as written in formula 3.14, $(L_S - N_W) - (L_N - N_{DI}) = N_{RD}$, is the complete expression for the relationship between sonar design parameters and the ocean sound transmission parameters. If this equation is rearranged to place all of the sonar design parameters on one side of the equality, the expression now becomes:

$$N_W = L_S - (L_N - N_{DI}) - N_{RD} \tag{3.15}$$

(Ocean Loss) = (Sonar Design Performance)

The summation of the foregoing sonar design performance parameters is called the sonar's figure of merit (*FOM*). Expression 3.15 can now be written:

$$N_W = FOM \tag{3.16}$$

Substituting from equation 3.4

$$20 \log R + a \times R \times 10^{-3} = \text{FOM} \tag{3.17}$$

Formula 3.17 now includes the range dependent terms on one side of the equality and the sonar dependent terms on the other. Considering figure of merit, one can see that by itself, it is a means of expressing quantitatively the design performance of a sonar. Figure of merit can be calculated by assuming a value of L_S and by measuring N_{DI}, L_N and N_{RD}. This calculated value of FOM is independent of range and is a basis for comparing the relative performance of two passive sonars *provided the calculations of the comparative figures of merit are made for the same frequency*. Another basis for comparing the design characteristics of two sonars is in terms of the Minimum Discernable Signal (*MDS*) detectable by the sonar. This calculation is the sum of L_N, N_{DI}, and N_{RD}, viz

$$MDS = L_N + N_{DI} - N_{RD} \tag{3.18}$$

Comparison of the MDS of one passive sonar with that of another at the same frequency is a meaningful comparison of the capabilities of the two passive sonars, and is independent of any assumption as to target radiated signal level, or of range.

The comparative tactical performance of two sonars is range dependent and may be made by first calculating the FOM's of the two systems concerned, and then solving for range (R), using equation 3.17 and a value for absorption loss (a) from figure 2-4.

Figure 3-2 is a plot of propagation loss vs. range showing spherical spreading loss only, and spreading loss plus absorption loss calculated from equation 3.4. Examination of figure 3-2 illustrates very clearly that even if two sonars, one at 5kHz and the other at 10 kHz, had the same numerical value for figure of merit, their tactical performance measured in detection range would be very different. Assume an FOM of 100 dB for each. The range prediction for the 10 kHz sonar is 16 kyds, while that for the 5 kHz sonar is about 32 kyd. For a third sonar, operating at a frequency of 1 kHz and having an FOM of 100 dB, the predicted range is 75 kyd. From a tactical point of view, for the propagation loss represented by figure 3-2, the lower frequency sonar has the greatest tactical capability, even though all sonars have the same FOM.

Suppose now that these three sonars had the following FOM's: 10 kHz – 105dB; 5kHz – 93 dB; and 1 Khz – 87 dB. Again referring to figure 3-2, it is seen that the range prediction for each sonar is 20 kyd, although the spread of values in FOM (allowable propagation loss) is 18 dB. From a tactical point of view, where the propagation loss is represented by the curves in figure 3-2, all of these three sonars are equal in performance.

From the discussion above, it might be concluded that comparing different sonars on the basis of figure of merit would not serve a useful purpose. This would not be entirely true. Consider three sonars of the same frequency (5 kHz) where their FOM's are 87, 93, and 105 dB, respectively. From figure 3-2, the ranges corresponding to these FOM's are as follows:

FOM (dB)	87	93	105
Range (kyd)	13	20	41

From this example, it is seen that for sonars operating at the same frequency, figures of merit are a valid and useful basis for comparing relative sonar performance because the difference in FOM's represent additional allowable propagation loss that can be sustained by the sonar having the higher FOM and still make a detection. The tactical value of this difference in FOM is readily expressed in units of range by using an appropriate propagation loss curve.

There is additional value in comparing the values of FOM for the same sonar taken over a period of time. A plot of such values taken at regular intervals over a period of time is a valuable indication of trends in sonar system condition, ship self-noise conditions, and operator proficiency.

Propagation Loss Curves

If all the world oceans had the same sound transmission characteristics, and propagation loss was subject only to the same spherical spreading loss and absorption loss in all parts of the ocean, figure 3-2 would be the only propagation loss curve needed to convert FOM to range. Measurements of sound transmission in many parts of the ocean have proven that not only does sound propagation change with frequency and location but with season as well. Therefore, in order to convert FOM to range, one must have a propagation loss curve for the frequency of the sonar(s) concerned and for the area and season of the year for which range prediction is desired. Figure 3-9 is a typical propagation loss curve for the waters off Iceland in the summer for a frequency of 2 kHz. Note that the measurements upon which this curve is based were made using a sound source and a receiver both at 50 ft., and both below the surface layer. Also note that propagation losses for three of the sound paths previously described are plotted. It is seen that in this area both multiple bottom bounce losses were measured and that multiple convergence zones were estimated.

Assume three different passive sonars with the following values of FOM at 2 kHz: 80, 95, 105. Based on figure 3-9, the range in kyds at which a 50% probability of detection is predicted for each sonar for each sound transmission path in the Iceland area is as follows (points labeled on the curve).

		Sound Path	
Sonar (FOM)	*Direct*	*Bottom Reflection*	*Convergence*
A 80	7.5	None	None
B 95	21	50	81
C 105	36	90	160

Figure 3-10 is a propagation loss curve for the Norwegian Sea area for the same frequency and season of the year as figure 3-9. Again, using sonars A, B, and C specified above, the predicted detection ranges in kyds. for this area are listed below.

		Sound Path	
Sonar (FOM)	*Direct*	*Bottom Reflection*	*Convergence*
A 80	5	None	None
B 95	7.5	30	60
C 105	11	56	170

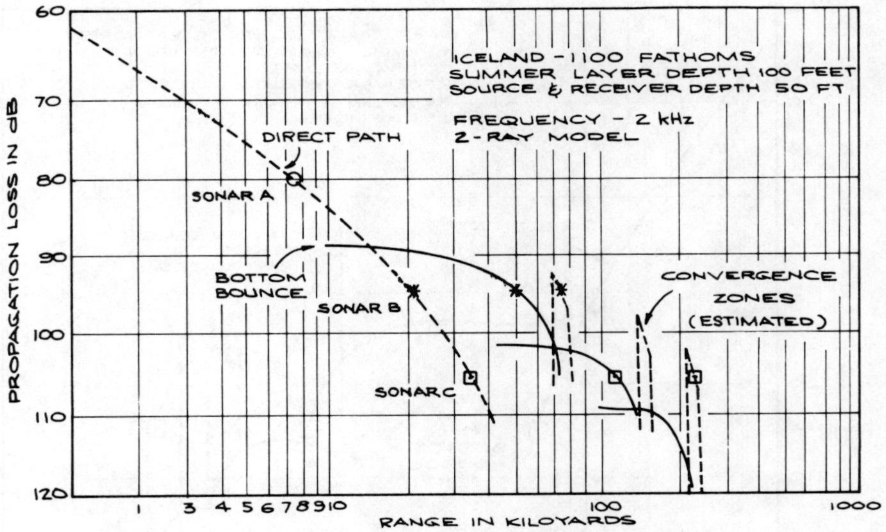

Source: Naval Operations Analysis, United States Naval Institute, 1968, fig. 9-8, p. 154. © 1968, United States Naval Institute.

Figure 3-9. Propagation Loss for the North Atlantic Ocean off Iceland in the Summer for 2 kHz

Combining the listings for these same sonars in both the Icelandic area and the Norwegian Sea illustrates the variability in tactical performance as measured by range in kyds for the same sonars in different ocean areas.

	Sound Paths/Ocean Area					
Sonar (FOM)	*Direct*		*Bottom Refl.*		*Convergence*	
	Ice.	*Norw.*	*Ice.*	*Norw.*	*Ice.*	*Norw.*
A 80	7.5	5	None	None	None	None
B 95	21	7.5	50	30	81	60
C 105	36	11	90	56	160	70

Note that in the direct sound path, sonar A with 80 dB FOM has the same tactical range prediction in the Icelandic area as does sonar B with 95 dB FOM in the Norwegian Sea. It can also be seen that the sonar with the higher figures of merit permit the use of sound paths (bottom bounce and convergence zone) not available to sonars with low figures of merit.

Figure 3-11 is an average sound velocity profile for two seasons

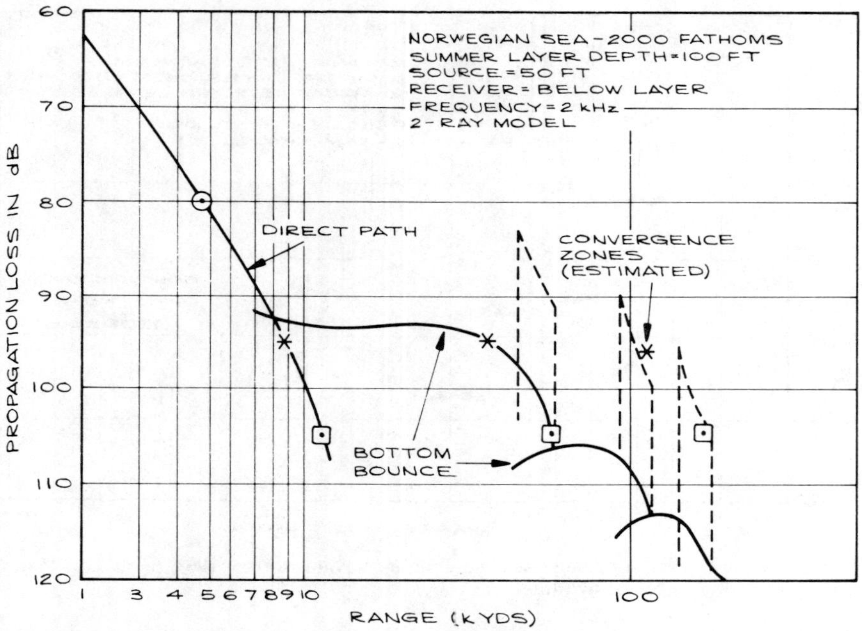

Source: Naval Operations Analysis, United States Naval Institute, 1968, fig. 9-9, p. 155. © 1968, United States Naval Institute.

Figure 3-10. Propagation Loss for the Norwegian Sea in the Summer for 2kHz

for an area in the North Atlantic east of Bermuda. Water depth in this area is about 14,500 ft. and the bottom loss is characterized as high according to Fleet Numerical Weather Center (FNWC) at Monterey, California, Figures 3-12 and 3-13 are the predicted propagation loss vs. range plots for this area based on these sound velocity profiles for a number of frequencies. These plots were obtained using a computer ray path program to convert the sound velocity profiles to propagation loss curves.

Figures 3-14, 3-15, and 3-16 are similar plots for the western Mediterranean Sea. Water depth in this area is about 8500 ft. and the bottom is characterized as low loss.

The principal reason for the difference in general appearance between these latter two sets of propagation loss curves and those of the Iceland and Norwegian Sea area is that the latter were made from a ray tracing program while the former were based on actual measurements and smoothed data. There is no significant difference in their use or application.

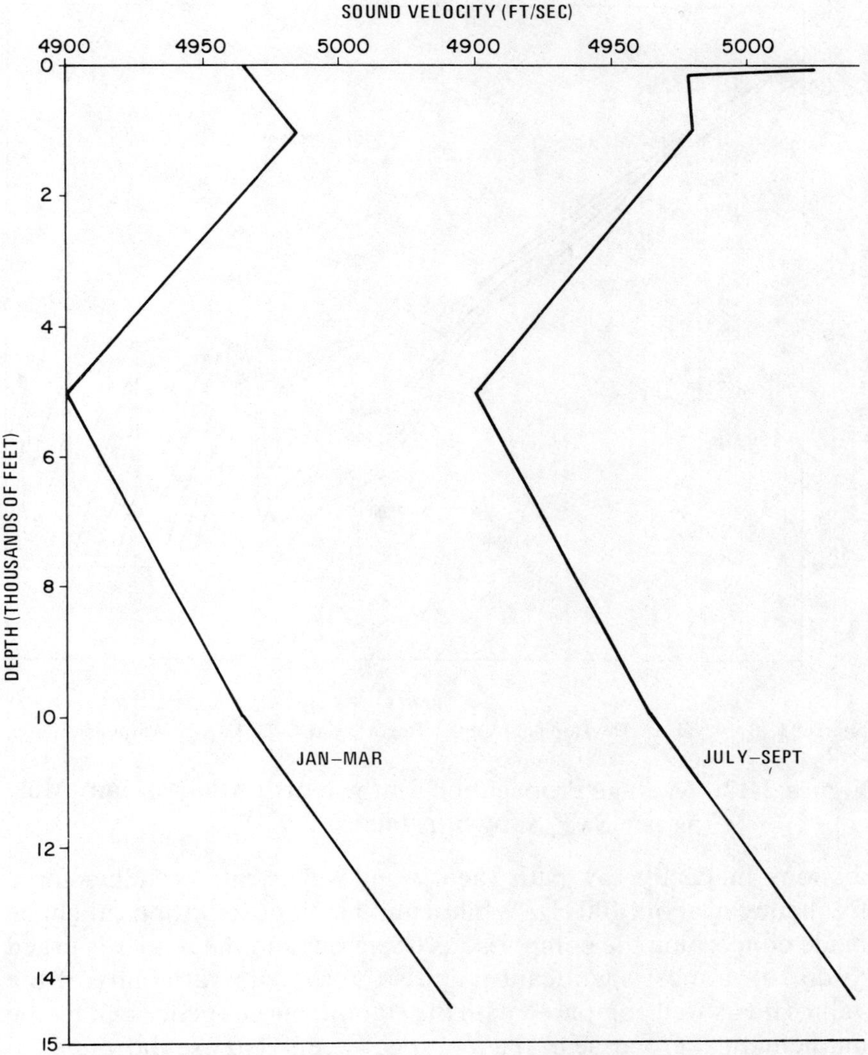

Source: U.S. Naval Air Development Report NADC-72012-AE, Warminister, Pa., 1972.

Figure 3-11. Average Sound Velocity Profiles for an Area in the North Atlantic East of Bermuda

One might well ask how well do ray path tracings compare with measured propagation loss data, and how well do both of these correlate with spherical spreading calculations? Figure 3-17 is a plot of propagation loss measured by personnel of the Naval Air Development Center, Warminister, Pennsylvania, compared with pre-

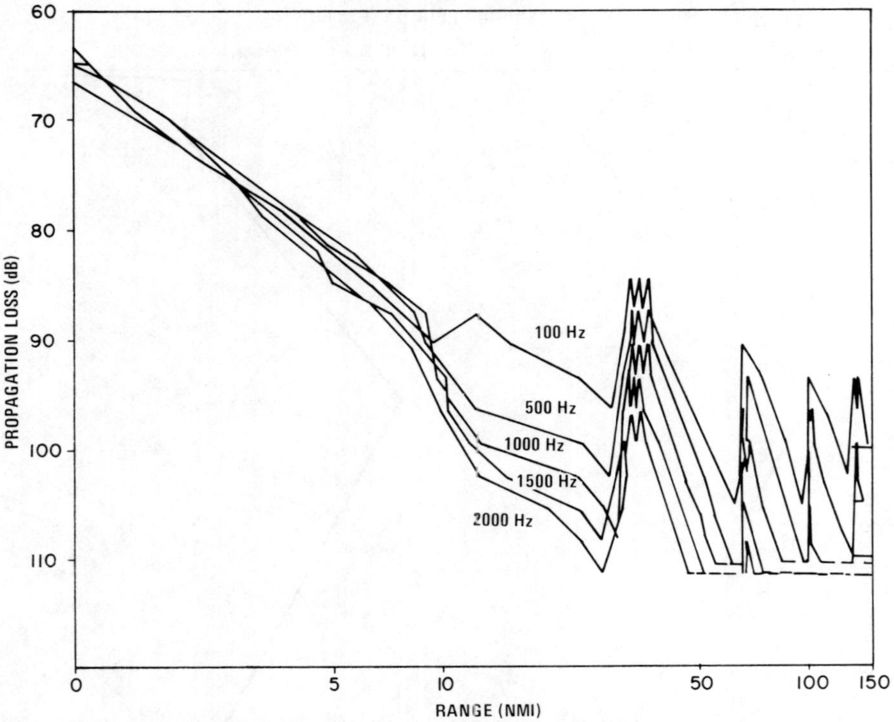

Source: U.S. Naval Air Development Center Report NADC-72012-AE, Warminster, Pa., 1972.

Figure 3-12. Average Propagation for the North Atlantic, Jan.-Mar. as per SVP Shown in Figure 3-11

dictions made by ray path theory and spherical spreading for a frequency of about 100 Hz. While a number of observations might be made concerning the comparisons depicted (and the reader is urged to do so), a most significant point is the random variability of the values measured compared with the smooth curves generated by the mathematical processes. The moral is, accept and use the values of the ray path plots as working tools in planning ASW operations, but be prepared for variation from the planning values as the rule rather than the exception. A corollary would seem to be, do not plan an operation where success or failure is dependent upon experiencing in nature a single value of propagation loss, but rather plan in terms of expecting to experience a spread of values. Just as there is no "standard ocean," neither is the writer aware of a universally accepted "standard deviation" for variation in ocean propagation

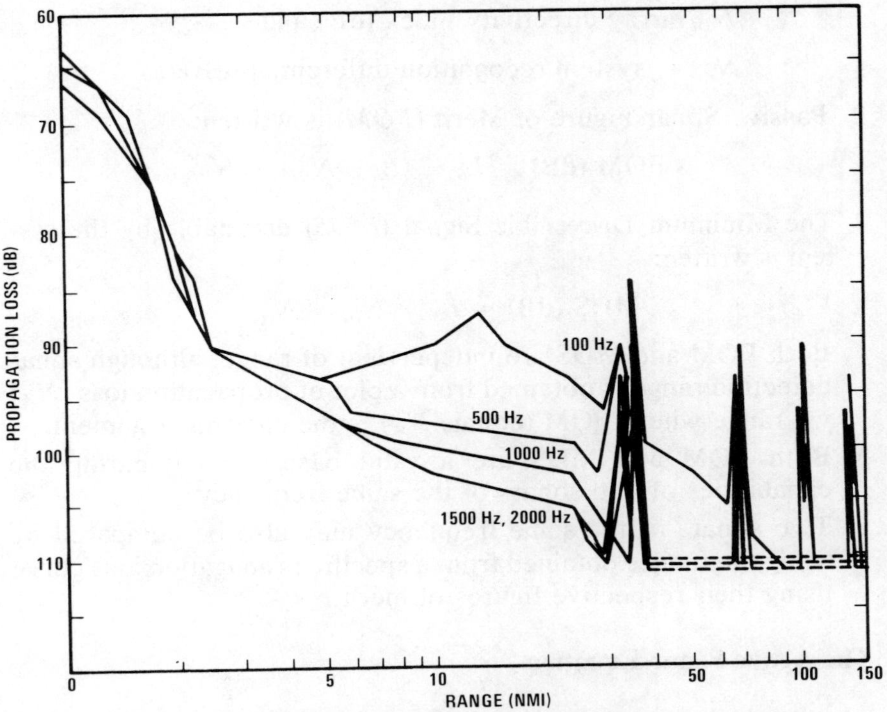

Source: U.S. Naval Air Development Center Report NADC-72012-AE, Warminster, Pa., 1972.

Figure 3-13. Average Propagation Loss for the North Atlantic, July-Sept. as per SVP Shown in Figure 3-11

losses. However, a number on the order of plus or minus 5-6 dB has some acceptance in the community as reasonable. A spread of detection ranges based on a variation in N_W of this order of magnitude should provide a reasonably low risk planning envelope.

Summary, The Passive Sonar Equation

1. The Passive Sonar equation is written:

$$N_W = L_S - (L_N - N_{DI}) - N_{RD},$$

where N_W is propagation loss (dB),

L_S is the target radiated noise signal (dB//(1 uPa)/(Hz)$^{\frac{1}{2}}$,

L_N is total background noise (dB//(1 uPa)/(Hz)$^{\frac{1}{2}}$

N_{DI} array directivity index (dB) and

N_{RD} is system recognition differential (dB).

2. Passive Sonar Figure of Merit (*FOM*) is written:

$$\text{FOM (dB)} = L_S - (L_N - N_{DI}) - N_{RD} .$$

3. The Minimum Discernible Signal (*MDS*) detectable by the system is written:

$$\text{MDS (dB)} = L_N + N_{DI} - N_{RD}.$$

4. Both FOM and MDS are independent of range, although sonar detection range is obtained from a plot of propagation loss (N_W) vs. range where FOM (equals N_W) is the entering argument.
5. Both FOM and MDS are a valid basis for comparing the capabilities of two sonars of the same frequency.
6. Two sonars of the same frequency may also be compared by using the ranges obtained from a specific propagation loss curve using their respective figures of merit.

The Active Sonar Equation

The active sonar equation is essentially the same as the passive sonar equation differing only in a few details. Consider the "signal" element of the equation. First, with an active sonar, the signal of interest L_S is now the radiated signal level of the sonar. Second, since the radiated energy from the active transducer must travel from the transducer to the target and back to the transducer, the propagation loss (N_W) is now twice that used in the passive situation. And third, since some of the output energy is reflected back from the target, a new term to represent the intensity of the echo from the target is introduced. Echo level or Target Strength is symbolized by N_{TS}. The signal portion of the active sonar equation is thus written:

$$\text{Signal Level} = L_S - 2N_W + N_{TS} . \qquad (3.19)$$

The noise term of the active sonar equation is written the same as for the passive equation, i.e.,

$$\text{Noise Level} = L_N - N_{DI}. \qquad (3.20)$$

The complete active sonar equation is written:

$$(L_S - 2N_W + N_{TS}) - (L_N - N_{DI}) = N_{RD}.$$

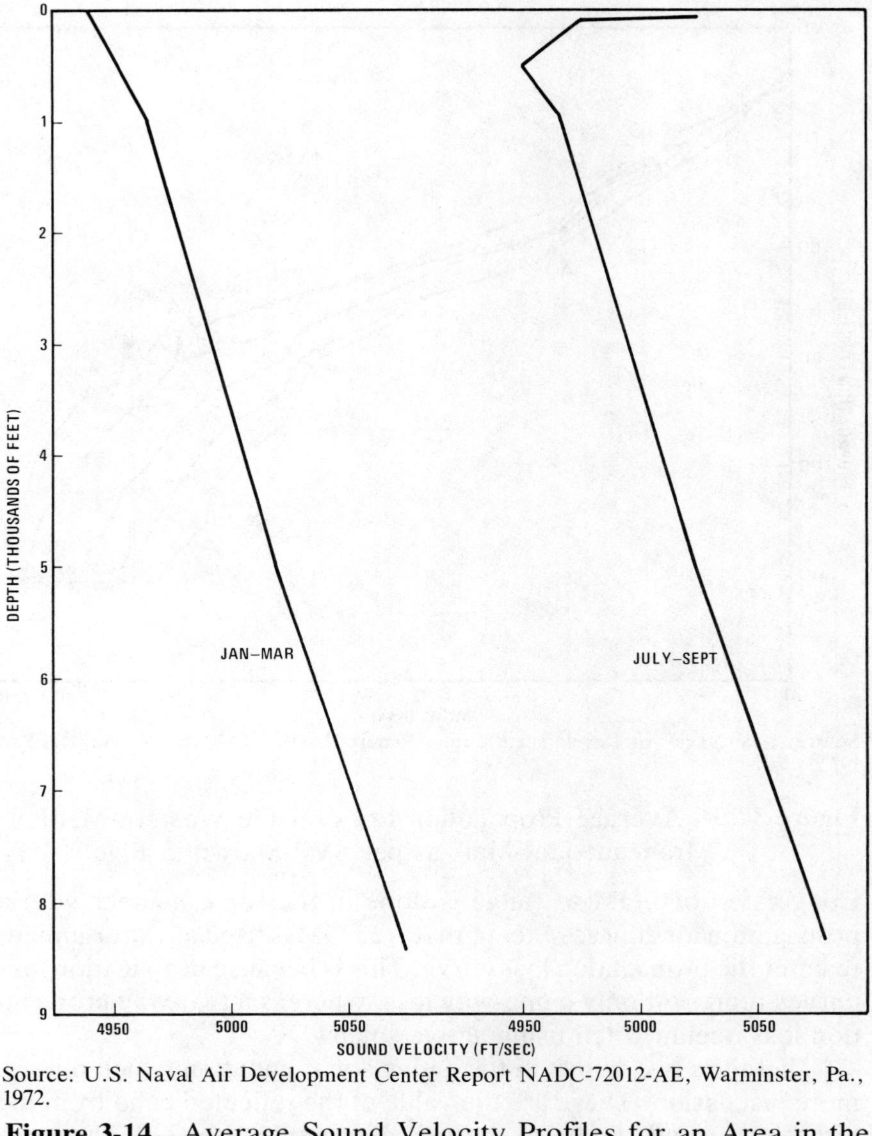

Source: U.S. Naval Air Development Center Report NADC-72012-AE, Warminster, Pa., 1972.

Figure 3-14. Average Sound Velocity Profiles for an Area in the Western Mediterranean

Transposing,

$$2N_W - (L_S + N_{TS}) - (L_N - N_{DI}) = N_{RD} \tag{3.21}$$

By definition, figure of merit (FOM) $=2N_W$, therefore

$$\text{FOM} = (L_S + N_{TS}) - (L_N - N_{DI}) - N_{RD}.$$

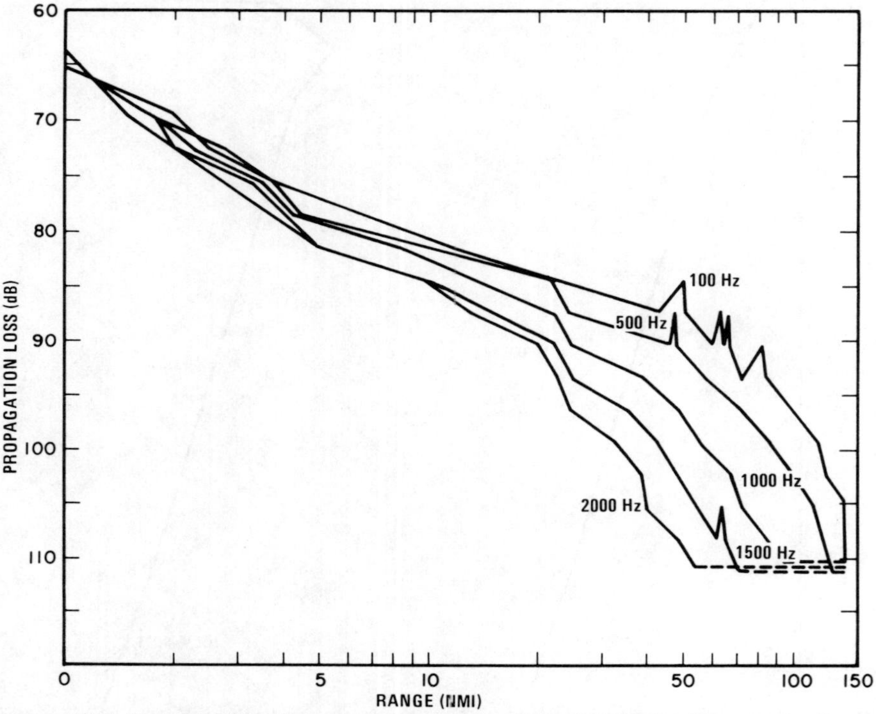

Source: U.S. Naval Air Development Center Report NADC-72012-AE, Warminster, Pa., 1972.

Figure 3-15. Average Propagation Loss for the Western Mediterranean, Jan.-Mar. as per SVP Shown in Figure 3-14

Conversion of FOM to range is done in the same manner with a propagation loss curve, except that 1/2 FOM is used as the argument to enter the propagation loss curve. This is because propagation loss curves represent only a one-way loss, whereas a two-way propagation loss occurs when using active sonars.

There are two points in the active sonar equation which require more discussion. These are the value of the reflected echo from the target (N_{TS}), and variations in the background noise (L_N). These are discussed below.

Target Strength (N_{TS})

The significant question to answer is "Why is N_{TS} always greater

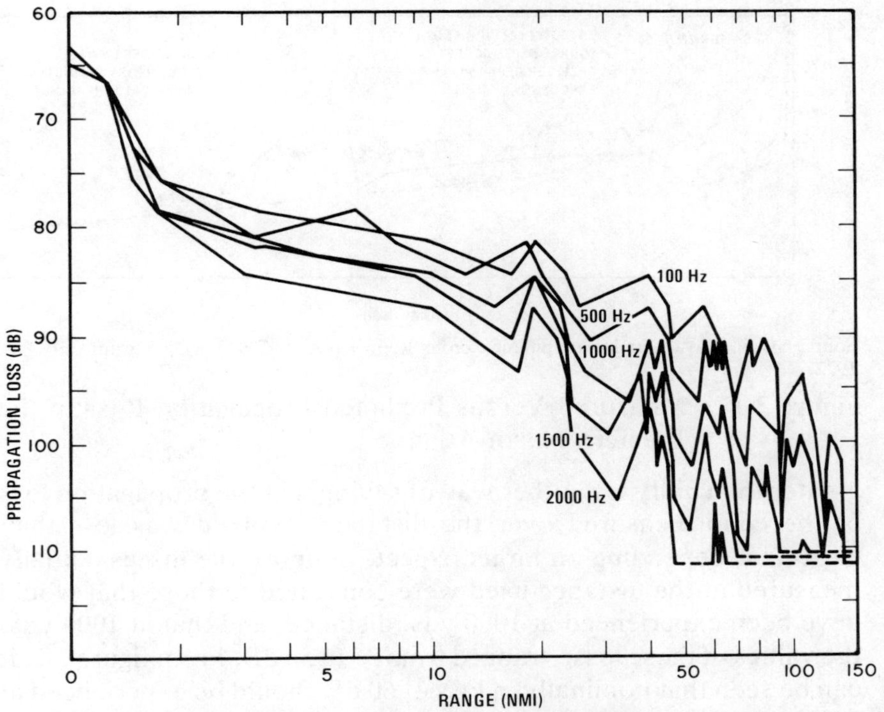

Source: U.S. Naval Air Development Center Report NADC-72012-AE, Warminster, Pa., 1972.

Figure 3-16. Average Propagation Loss for the Western Mediterranean, July-Sept. as per SVP Shown in Figure 3-14

than unity?" A secondary question is "Does it vary, and if so, how?"

Values of N_{TS} normally range from 10 dB to 25 dB. That these values are average values observed by measurement at sea is the simple and straightforward answer to the first question above. The variability of the value of N_{TS} (from 10-25 dB) is a function of the target aspect presented to the incoming signal. A beam target presents a greater reflective area than the bow, hence more energy is reflected from a beam target (25 dB) than from a bow target (10 dB). In calculating figure of merit, a value of 15 dB is considered a nominal average value of N_{TS}.

The "why" of the positive value of N_{TS} is worth a few words more. As stated above, the values cited resulted from measurements made at some distance from the reflecting target. Stating that N_{TS} is

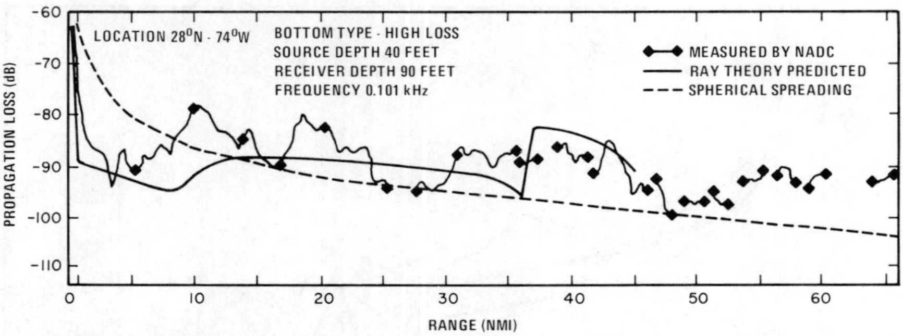

Source: U.S. Naval Air Development Center Report NADC-72012-AE, Warminster, Pa., 1972.

Figure 3-17. Measured Versus Predicted Propagation Loss in the Western North Atlantic

greater than unity is another way of saying that the propagation loss of the echo measured over the distance involved was less than expected, depending on target aspect. Suppose the losses actually measured at the distance used were converted to those that would have been experienced at 1000 yds. distance, and that at 1000 yds. the values of these losses ranged from 35 to 50 dB. From figure 3-2, it can be seen that nominally, a loss of 60 dB should be experienced at 1000 yds. It is now apparent that the difference between the expected and measured values of propagation loss is readily handled by assigning this difference to N_{TS} as a positive value.

Why should the loss in this instance be less than normal? Theory has it that for large sound sources (such as the hull of a submarine), the large source can be considered to consist of a number of small elemental areas, each of which can be considered to act as a point source of sound, radiating in all directions. Theory also states that "if the linear dimensions of the source (point) are small compared to the wavelength of the sound, the difference in distances from a remote point in the sound field to two elemental areas on the source are small compared to the wavelength,"[1] so that the sound pressures from all of the small point sources arrive at the remote point essentially at the same time, and therefore are additive at any given point.

Background Noise Level for Active Sonars

In the case of the passive sonar situation, the background noise

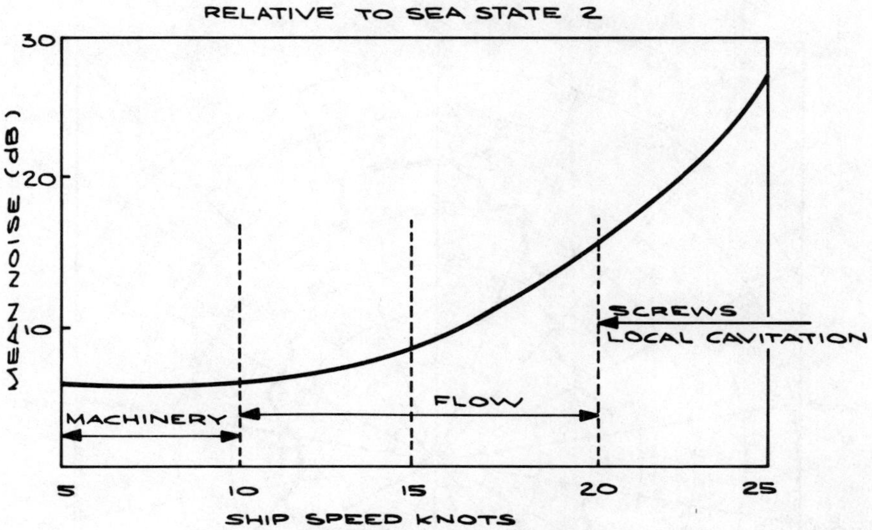

Source: *Introduction to Sonar Technology,* Bureau of Ships, Navy Department, Washington, D.C., fig. 4-7, p. 92.

Figure 3-18. Typical Self-Noise Behavior (LN Component) for a Destroyer

elements were either the ambient noise of the sea, or self-noise from the ship, which included both flow noise over the hydrophones and ship machinery-generated noise. Ambient noise levels were depicted in figure 3-3, and it was indicated that usually passive sonar platforms operated at self-noise levels such that sea noise was the predominant component of L_N.

Figure 3-18 is a plot of typical self-noise levels for a destroyer relative to sea state 2. From this it can be seen that for an echo ranging destroyer operating at nominal screening speeds, ship self-noise will be the predominant component of L_N as between ambient and ship self-noise. Figure 3-19 is a plot of the directionality of self-noise for a typical destroyer.

For active sonars, there is another type of noise for consideration—reverberation. Reverberation was defined in Chapter 2, under "Scattering and Reverberation." The net effect of this echoing and re-echoing of the active sonar's transmitted pulse is to create a level of sound reverberation in the sea that at times will be greater than ships self-noise levels dependent upon the power levels used and the quantity of scatterers present.

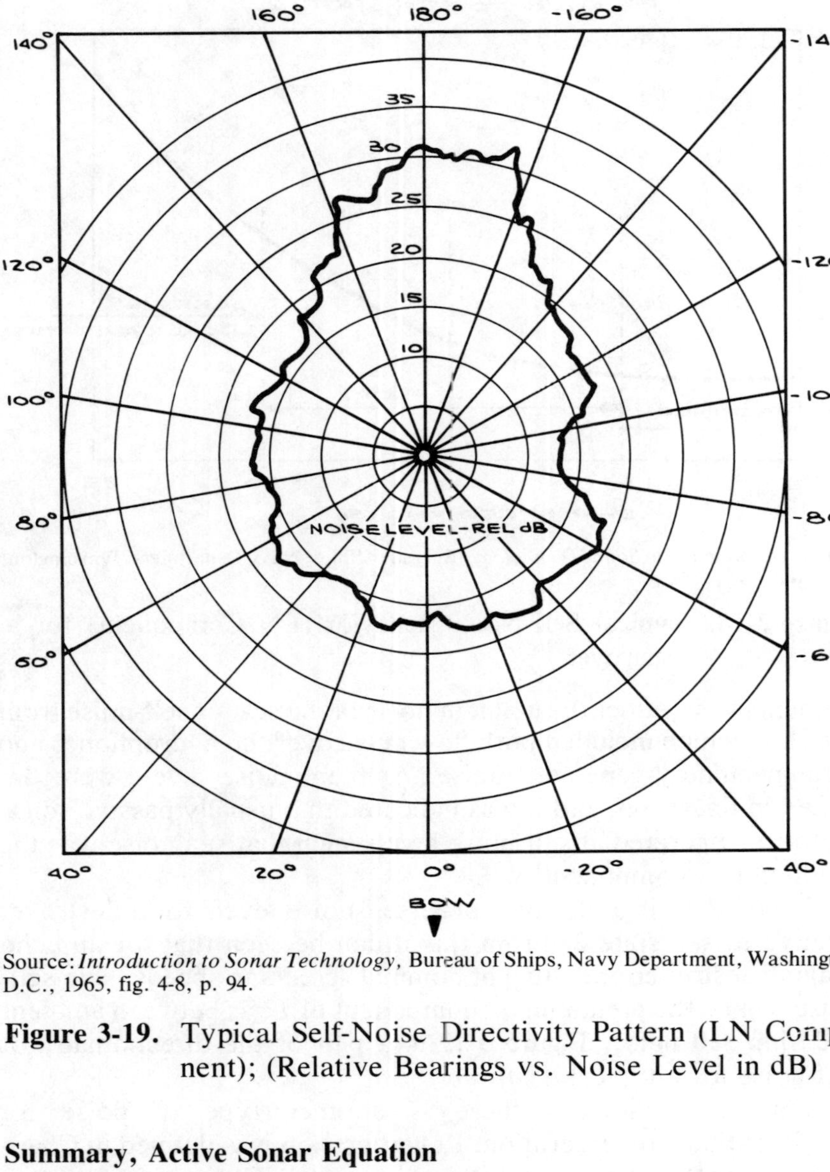

Source: *Introduction to Sonar Technology,* Bureau of Ships, Navy Department, Washington, D.C., 1965, fig. 4-8, p. 94.

Figure 3-19. Typical Self-Noise Directivity Pattern (LN Component); (Relative Bearings vs. Noise Level in dB)

Summary, Active Sonar Equation

1. The active sonar equation is written $(L_S - 2N_W + N_{TS}) - (L_N - N_{DI}) = N_{RD}$, where N_W, L_N, N_{DI}, and N_{RD} are as described above in "Summary, The Passive Sonar Equation," under #1. L_S is the

sonar transmitted source level (dB//(1 uPa)/(Hz)$^{\frac{1}{2}}$, and N_{TS} is target strength in dB.

2. Active sonar figure of merit (*FOM*) is written $FOM\,(\text{dB}) = (L_S + N_{TS}) - (L_N - N_{DI}) - N_{RD}$.
3. The minimum discernable signal (*MDS*) detectable by the system is written $MDS\,(\text{dB}) = L_N + N_{DI} - N_{RD}$.
4. Both FOM and MDS are independent of range, although detection range is obtained from a plot of propagation loss (N_W) vs. range where 1/2 *FOM* (equals N_W) is the entering argument.
5. FOM is a valid basis for comparing the capabilities of two active sonars operating at the same frequency.
6. Two active sonars of the same frequency may also be compared by using the ranges obtained from a specific propagation loss curve using their respective figures of merit.

The Measurement of Figure of Merit at Sea

The figure of merit expressions for both passive and active sonars are restated for convenience.

Passive Case

$$FOM = L_S - (L_N - N_{DI}) - N_{RD}.$$

Active Case

$$FOM = (L_S + N_{TS}) - (L_N - N_{DI}) - N_{RD}.$$

Figure of merit can be obtained quantitatively by

a. measuring or assuming a value of L_S, as appropriate to the type sonar;
b. measuring N_{DI}, L_N, and N_{RD}; and
c. assuming a value for N_{TS} (usually 15 dB).

N_{DI} is a function of physical hydrophone spacing in the sonar array and is fixed by design and installation. As long as the hydrophone or transducer elements in the array are in operation, the N_{DI} as measured at the time of construction of the array (or in a prototype array) is valid for use. There is no need to attempt to measure this sensor parameter at sea.

It is possible to measure L_N and N_{RD} combined. It is also possible to measure L_N as a separate quantity and then using prescribed calibration curves which assume an average value of N_{RD}, to combine these data with L_S and N_{DI} to obtain FOM. A general description of making these measurements is indicated below. The important point to be made here is that these quantities can be measured, and should be measured, on a regular basis to determine if ship machinery noise levels have increased with time and wear, or to determine if operator perception may have changed with time.

For active sonars, an additional measurement to be made is sonar source level. This can be measured by observing the electrical power output in watts and converting this to decibels at the output of the transducer. Target strength in the case of active sonars is usually established at an average value of 15 dB.

As stated earlier, it is possible to measure L_N and N_{RD} combined. It is also possible to measure L_N as a separate quantity and then using prescribed calibration curves which assume an average value of N_{RD}, to combine these data with L_S and N_{DI} to obtain FOM. First, to measure L_N separately, assume a ship has arrived at its assigned operating area, and it is desired to measure FOM. The vessel takes search speed on the assigned or expected course. All machinery ordinarily in operation is in use. The sonar is manned, and in the passive mode, the reading on the signal level meter of the sensor is recorded every ten degrees of bearing. A composite or average of the levels may be used, or the value for a single bearing used as L_N. Note that under these conditions, the quantity indicated on the meter includes actual ocean ambient and vessel flow noise at the sonar array plus nearfield ship noise as heard by the sonar hydrophones. As previously stated, the value selected is used to enter a calibration curve, and FOM can be read out.

To measure L_N and N_{RD} combined, place the ship in the same posture as before. This time, a signal of known level is superimposed on the L_N quantity one sees through the hydrophones as described above. This signal, injected into the system at the preamplifiers just *inboard* from the hydrophones is varied until the operator just detects the target energy signature for passive sonars or the synthetic echo in the case of an active sonar. The level of the signal required to be just discernible to the operator is the sum of L_N and N_{RD}. Note that to perform this measurement, a calibrated signal source is required. Also note that this measurement includes the performance of the

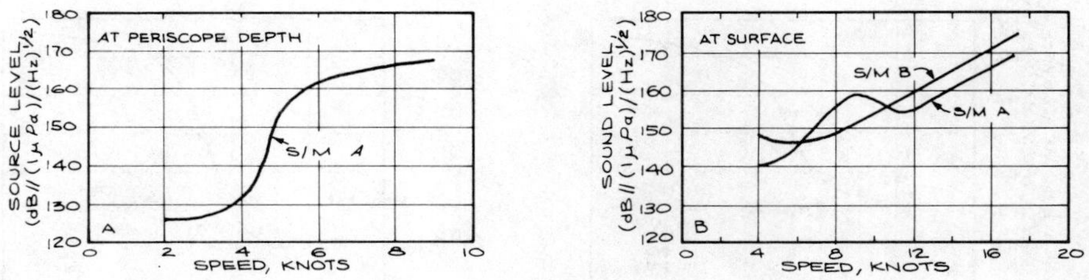

Source: National Defense Research Committee (NDRC), *Principles and Applications of Underwater Sound* (Washington, D.C.: U.S. Government Printing Office, 1946), fig. 9, p. 232.

Figure 3-20. Overall Source Levels of Submarine Sounds. (A) Submerged Variation with Speed; (B) Two Submarines, Surface Operation, Illustrating the Variability Between Ships

sonar operator as well as that of the equipment. And a final note, recognize that in the case of the passive sonar, the extent to which the synthesized target signature varies from a true target signature may bias the measurement. This method requires a target signature or target echo generator.

Some typical average values for submarine and ship radiated noise levels (L_S) are depicted in figures 3-20, 3-21, 3-22, 3-23.

If figure of merit can be measured while underway at sea, the commanding officer of a ship can do these important things:

1. Determine that his sonar is or is not operating as designed.
2. Predict a detection or localization range for various targets *if* propagation loss versus range curves are available.

With the figure of merit concept in mind, the war planner can do these things:

1. Compare the capabilities of various sonars for various ships, both real and hypothetical.
2. Predict the detection ranges possible in various areas *if* propagation versus range curves are available.

Machine-aided Performance Prediction

The advent of the digital computer along with computer-driven graphic television screen type displays, with which some ships are

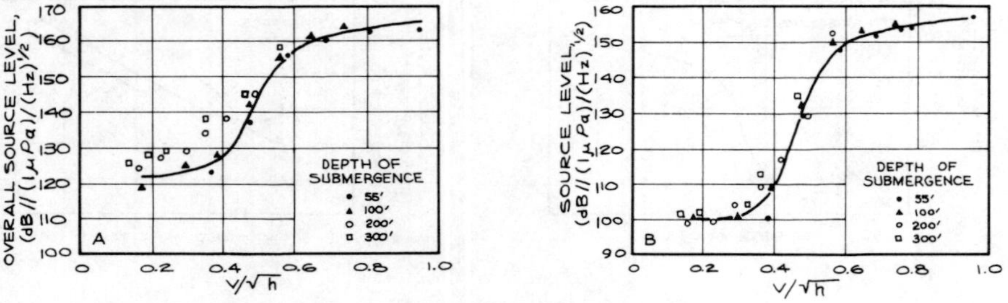

Source: National Defense Research Committee (NDRC), *Principals and Applications of Underwater Sound* (Washington, D.C.: U.S. Government Printing Office. 1946), fig. 10, p. 233.

Figure 3-21. Dependence of Overall Source Levels of Submarine Sounds on Depth of Submergence h(feet) and speed V(knots). The steep rise between $V/h^{\frac{1}{2}} = 0.4$ and $V/h^{\frac{1}{2}} = 0.6$ is due to cavitation. The solid curve is drawn on the assumption that the speed at which cavitation occurs is inversely proportional to the square root of the hydrostatic pressure. Figure A plots the levels measured in the 0.1- to 10-kc bandwidth: Figure B the levels in the 10- to 30-kc band.

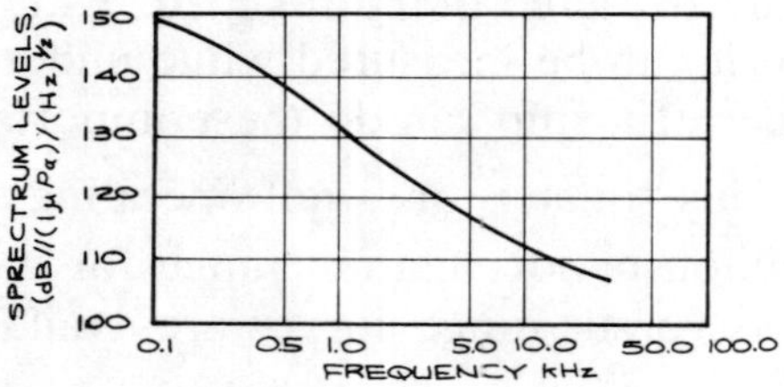

Source: National Defense Research Committee (NDRC), *Principals and Applications of Underwater Sound* (Washington, D.C.: U.S. Government Printing Office, 1946), fig. 11, p. 233.

Figure 3-22. Average Spectrum of a Submarine Running at 6 kts at Periscope Depth or 12 kts at the Surface. It must be stressed that the spectra of individual ships may deviate decidedly from this figure.

being equipped, provides a powerful new tool to the ASW tactical commander, particularly at the ship level. These display systems are programmed to accept the temperature or sound velocity profile measured, and to display the sound velocity profile to scale on the

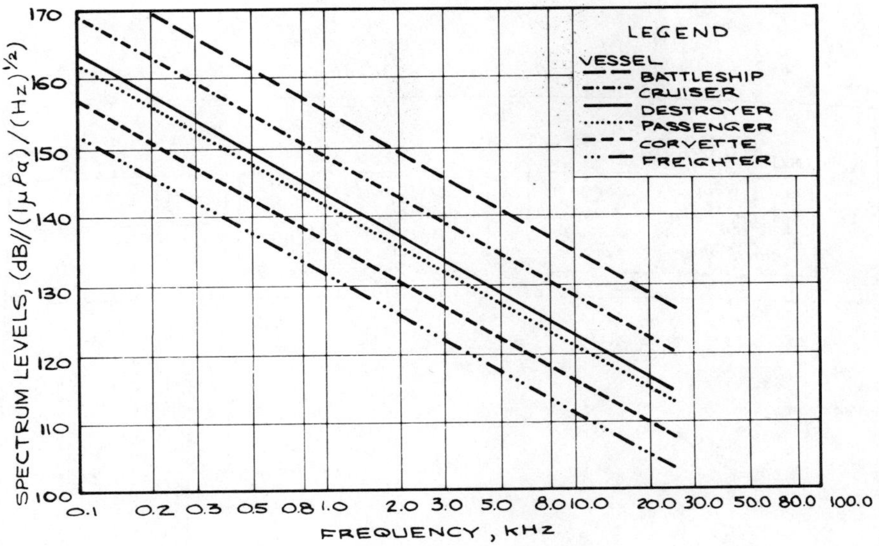

Source: *Introduction to Sonar Technology,* Bureau of Ships, Navy Department, Washington, D.C., 1965, fig. 2-3, p. 33.

Figure 3-23. Average Radiated Noise Levels of Six Different Classes of Ships (L_S)

screen. Figure 3-24 is a photograph of one such display. Note that the specific area where the data are taken is displayed, as well as the time of the measurement in the upper left hand box. Also, note that actual temperature values are displayed in the table to the right of the graphic sound velocity profile. This table assists in determining the depth at which layers, or marked changes in rate or direction of variation of sound velocity occur. While figure 3-24 shows the profile from surface to bottom, it is possible to examine a segment of the sound velocity profile in more detail by changing scale. Figure 3-25 is a photograph of the top 3000 ft. of water with a different sound velocity profile.

After the SVP has been entered, displayed, and examined, the sonar figure of merit is added, along with certain of the operating parameters of the sonar such as transducer depth and beamwidth of the active sonar transducer. On the basis of this additional data, the machine then computes and displays the resulting ray path geometry of the sonar transmissions according to the sound velocity profile existing. (These are also the sound paths for a sound source such as a

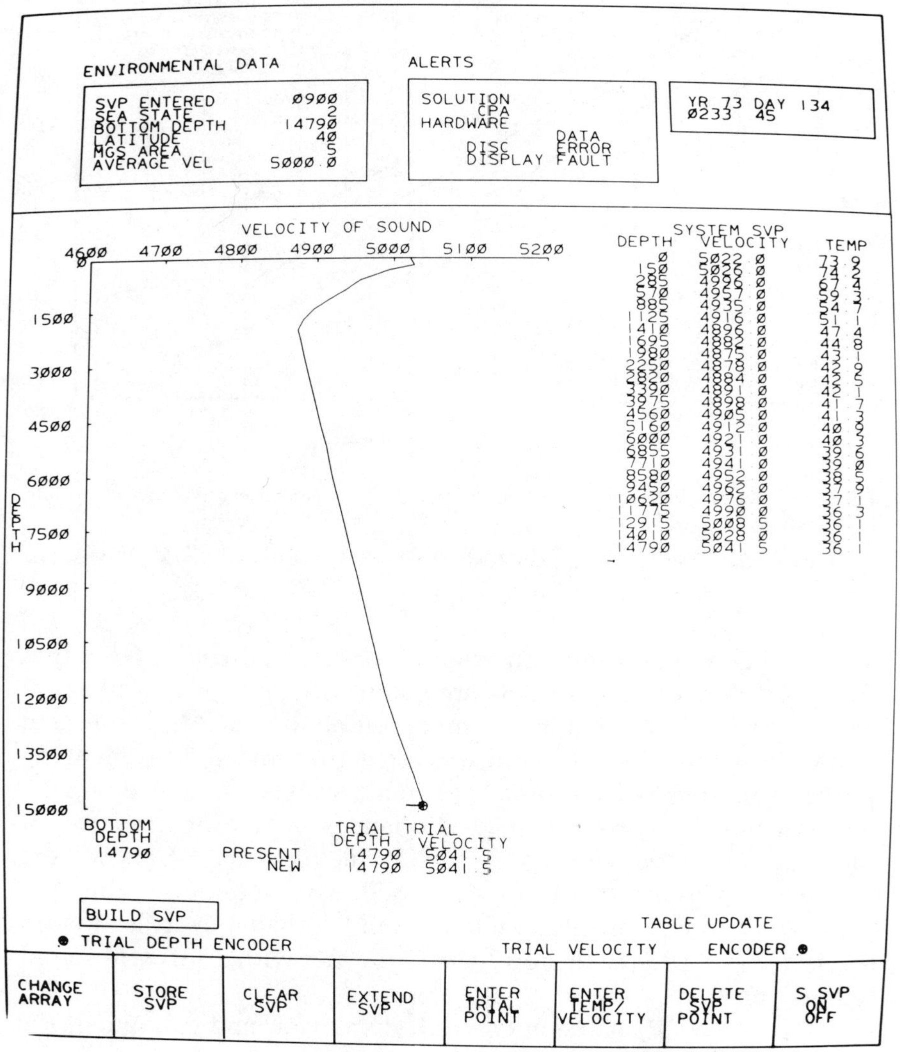

Source: Raytheon Company Photograph.

Figure 3-24. Photograph of Raytheon *M*ulti- *P*urpose *S*ubsystem (MPS) Display Showing Sound Velocity Profile Trace

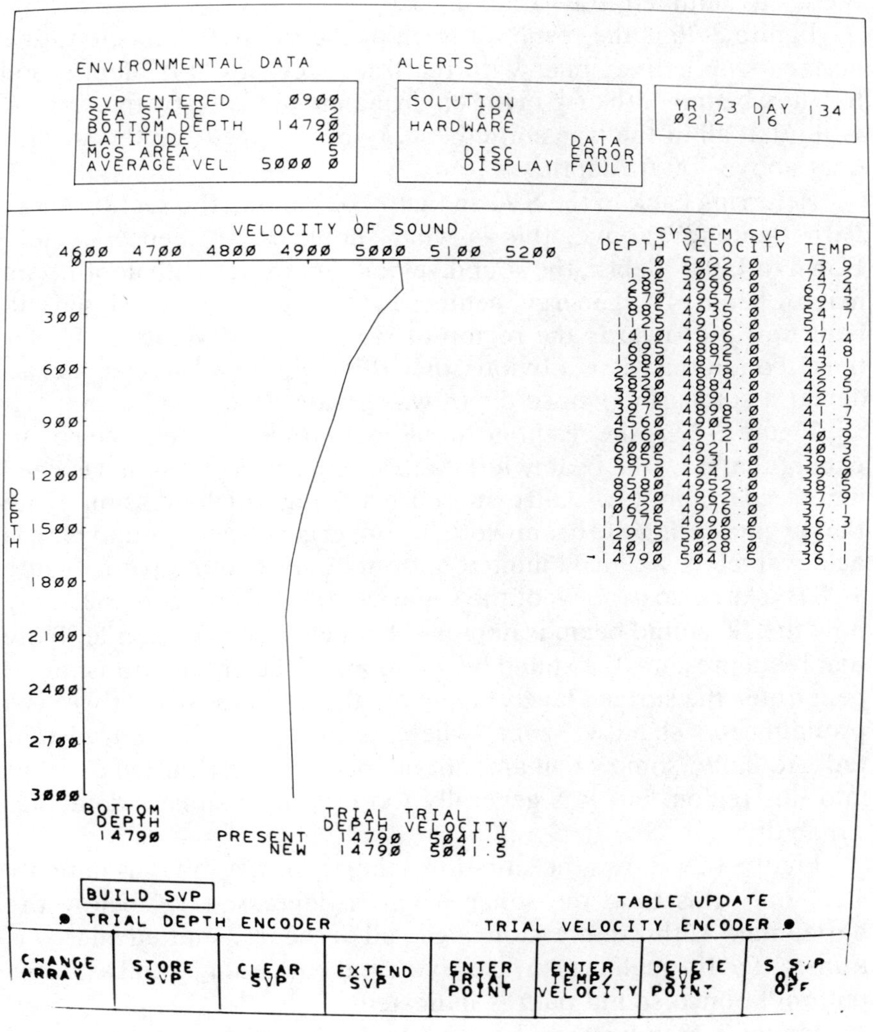

Source: Raytheon Company Photograph.

Figure 3-25. Raytheon MPS Display Photograph Showing a Different Sound Velocity Profile Trace

ship or submarine radiating its acoustic signature at that depth.) These ray paths are plotted generally in accordance with ray path theory as outlined in Chapter 2.

Figure 3-26 is the resultant trace of the ray path expected for a surface ship active sonar with the transducer at 20 ft. depth, and having a beamwidth of 4° in the environment depicted by figure 3-24. Note that all of the transmitted energy is "trapped" in the surface duct above 150 ft. depth.

Referring back to the SVP in figure 3-24 and to the section "Ray Path Bending" above, this is what should have been expected. Down to 150 ft. depth, the sound velocity increases with depth. The narrow beam (4°) of energy, centered at 0° angle of depression, is all bent upward towards the region of slower sound velocity. Under these conditions, it is obvious that this ship could not expect to detect a submarine whose depth was greater than 150 ft.

Figure 3-27 is the resultant sound path trace for the same sonar, except that now the beamwidth of that sonar transmission has been increased to about 12°, still centered at a 0° angle of depression. Note that by increasing the beamwidth, a convergence zone sound path is achievable, as well as a limited bottom bounce sound path. Figure 3-28 is a large-scale view of the same situation as in figure 3-27 only now the 12° sound beam is depressed 5° below the horizontal. Note that by depressing the sound beam, more of the energy transmitted penetrates the surface layer. Also note that in both cases, there is a pronounced "shadow" zone, where no energy is shown transmitted. Actually, some small amount of energy is transmitted or leaks into this region, but it is generally too little to assume a detection capability.

Figure 3-29 shows the situation using the same sonar as in figure 3-28, only this time the sonar beam is depressed 15° below the horizontal. At this depression angle, all of the transmitted energy is transmitted through the surface layer, and both a single and a double bottom bounce sound path is indicated.

Figure 3-30 indicates a large-scale display of the paths taken by the radiated noise signature of a submarine or other sound source at a depth of 500 ft. in the environment in figure 3-30. While it is recognized that the radiated noise signature of a ship is basically omnidirectional, it is useful on this display to show only about 17 rays spaced 2° apart, on a 32° beam.

And finally, figure 3-31 depicts the resulting acoustic geometry

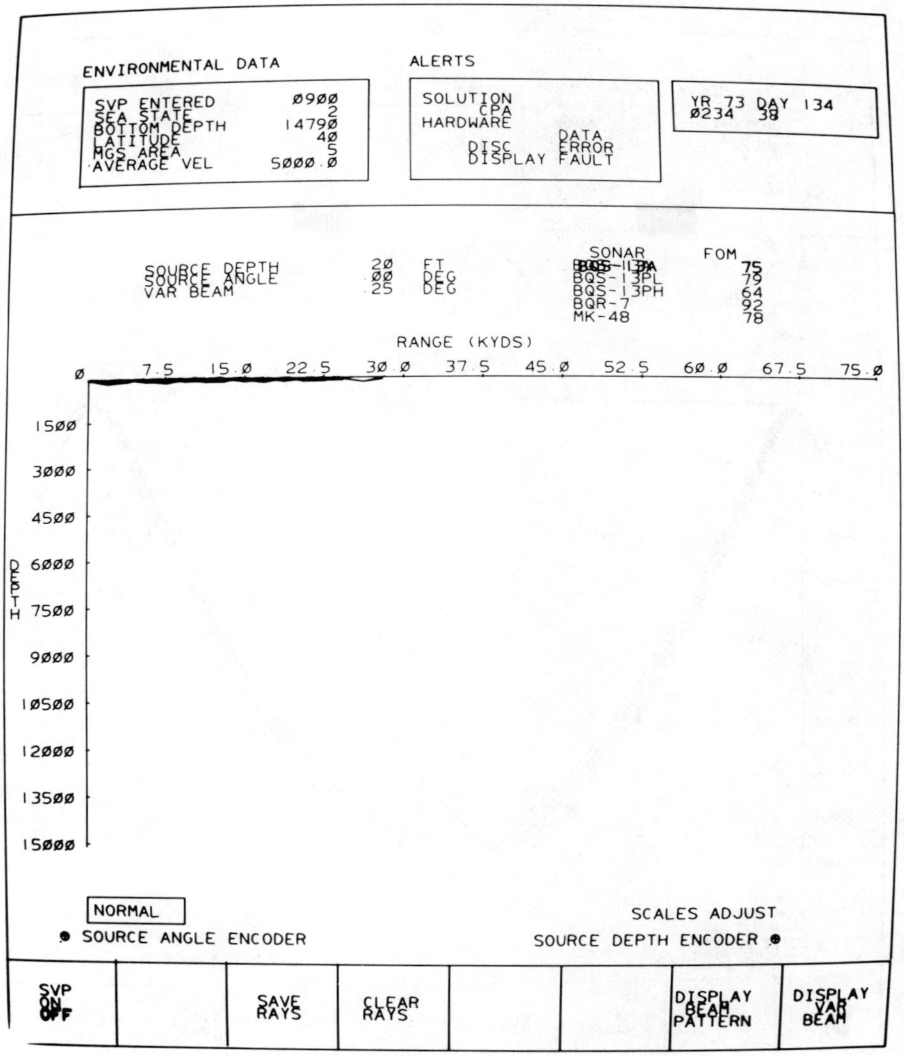

Source: Raytheon Company Photograph.

Figure 3-26. Raytheon MPS Display Photo Showing Ray Trace for the Conditions Noted Above the Ray Trace

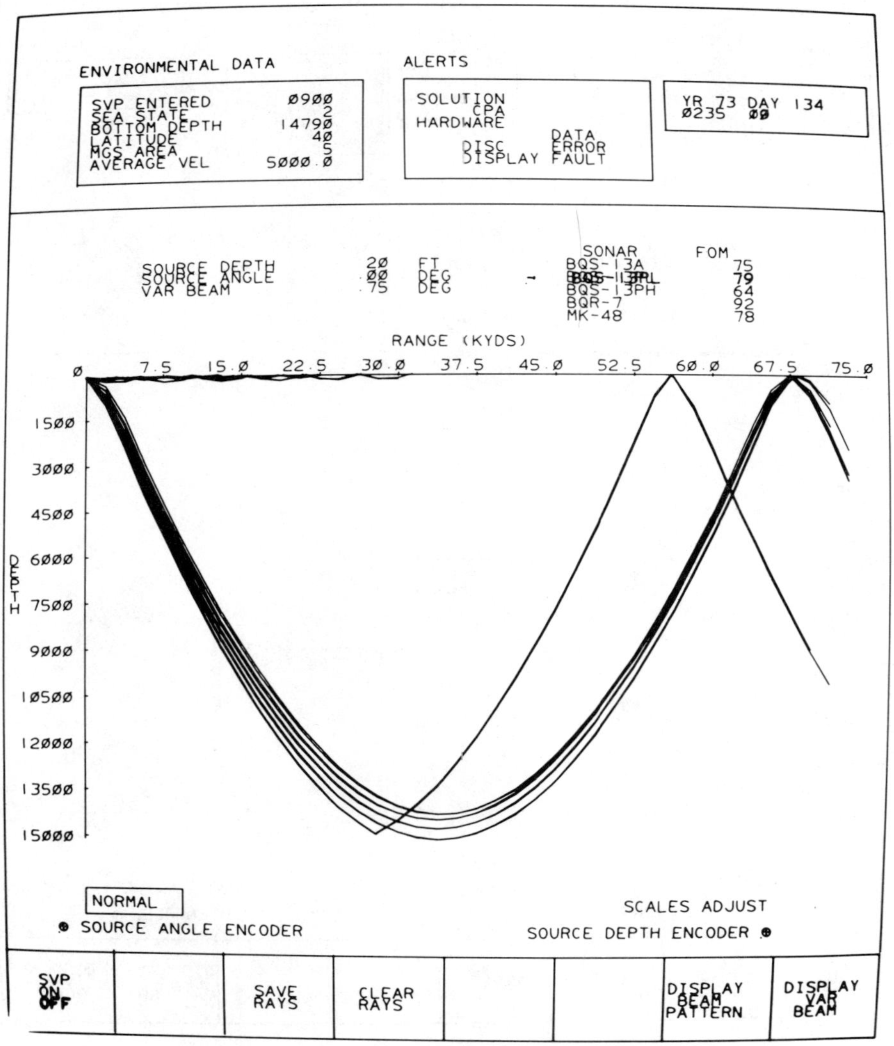

Source: Raytheon Company Photograph.

Figure 3-27. A Similar Ray Path Trace on the Raytheon MPS Display, Showing a Surface Duct and Convergence Zone Sound Paths

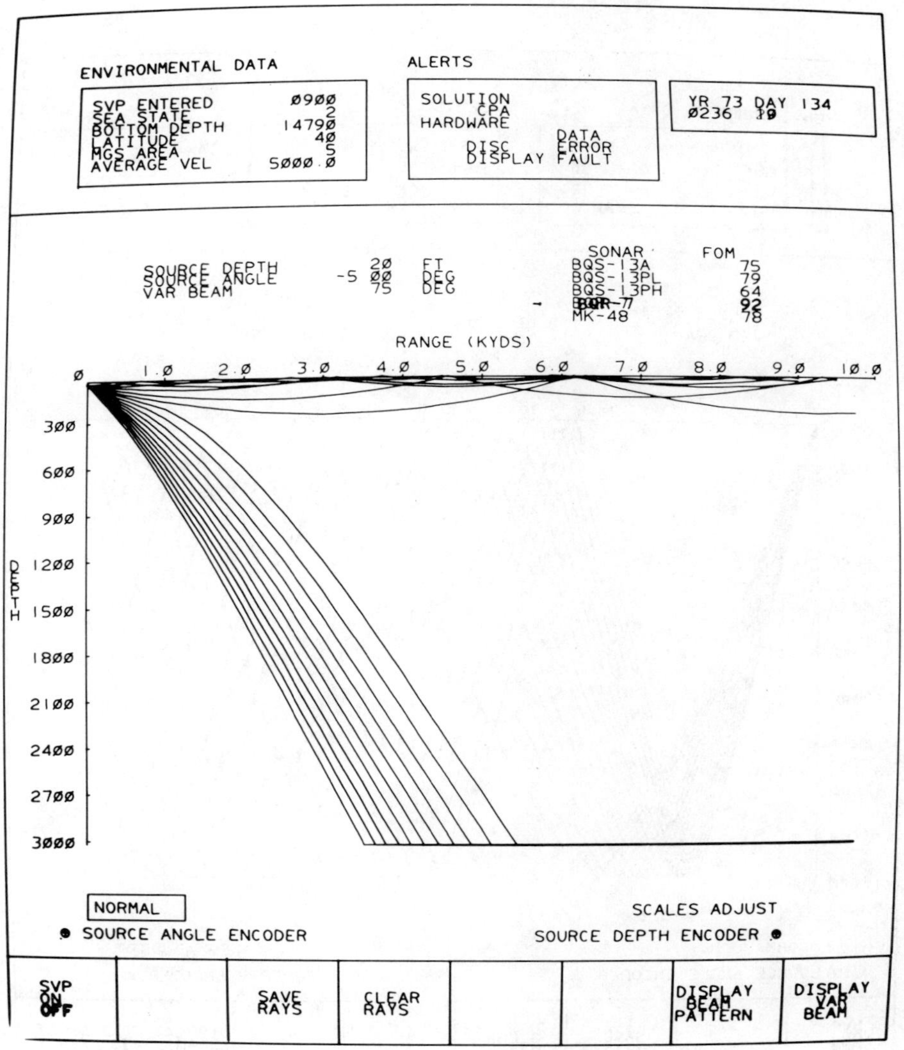

Source: Raytheon Company Photograph.

Figure 3-28. An Expanded Ray Trace of the Upper 3000 Ft. of the Trace Shown in Figure 3-27

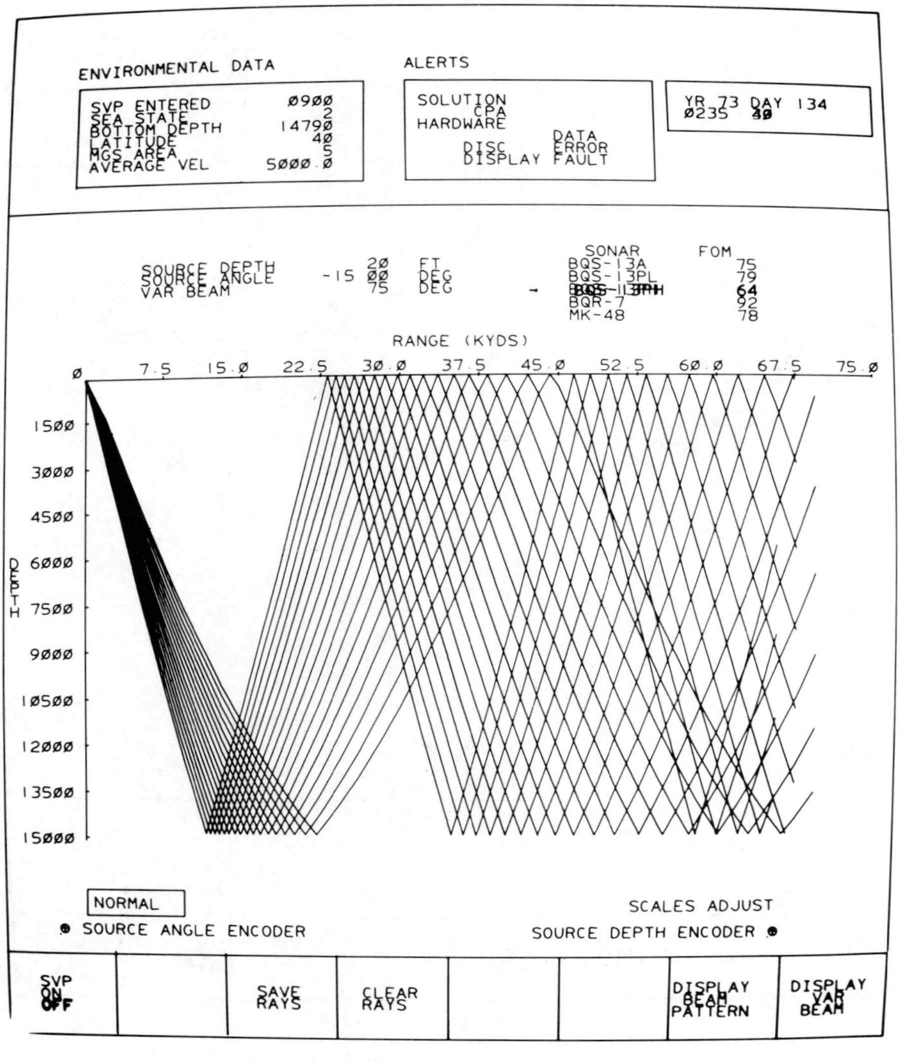

Source: Raytheon Company Photograph.

Figure 3-29. A Similar Ray Trace on the Raytheon MPS Display, Showing Multiple Bottom Reflected Sound Paths

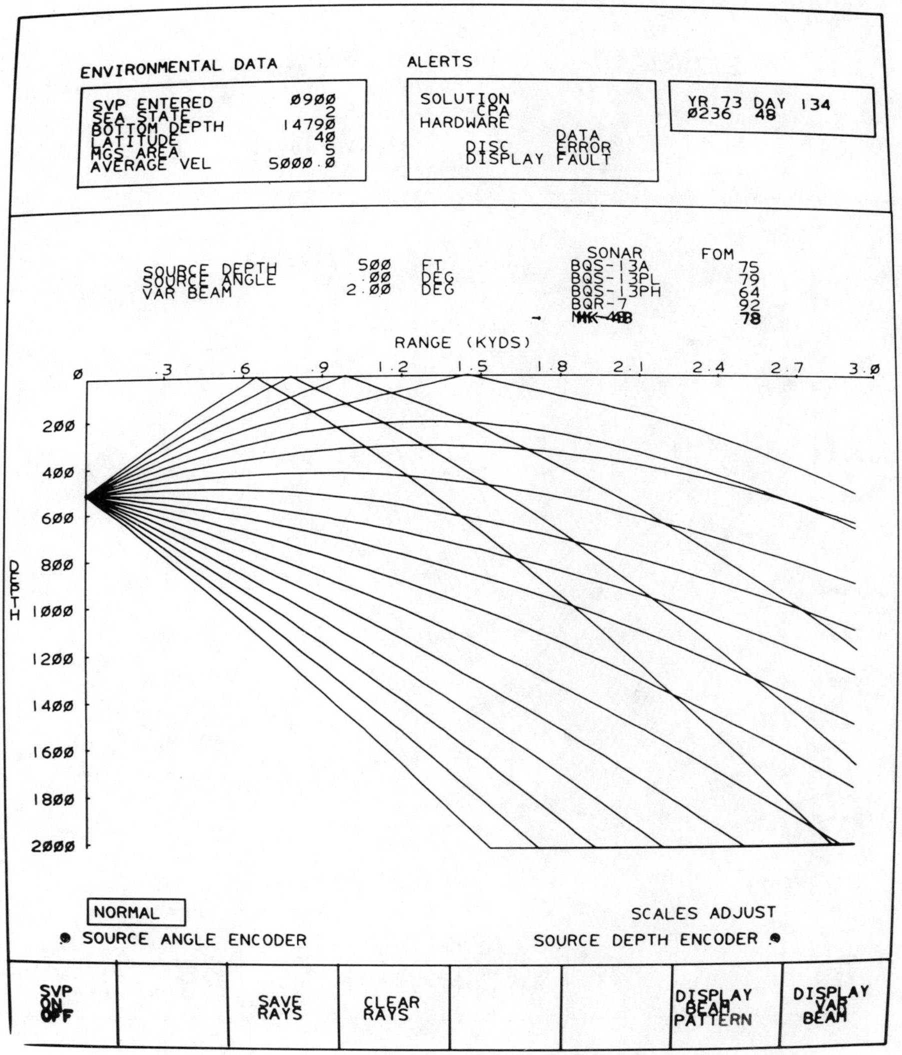

Source: Raytheon Company Photograph.

Figure 3-30. A Large-Scale Trace on the Raytheon MPS Display Showing the Sound Paths Taken by the Radiated Noise Signature of a Submarine at 500 Ft. Depth

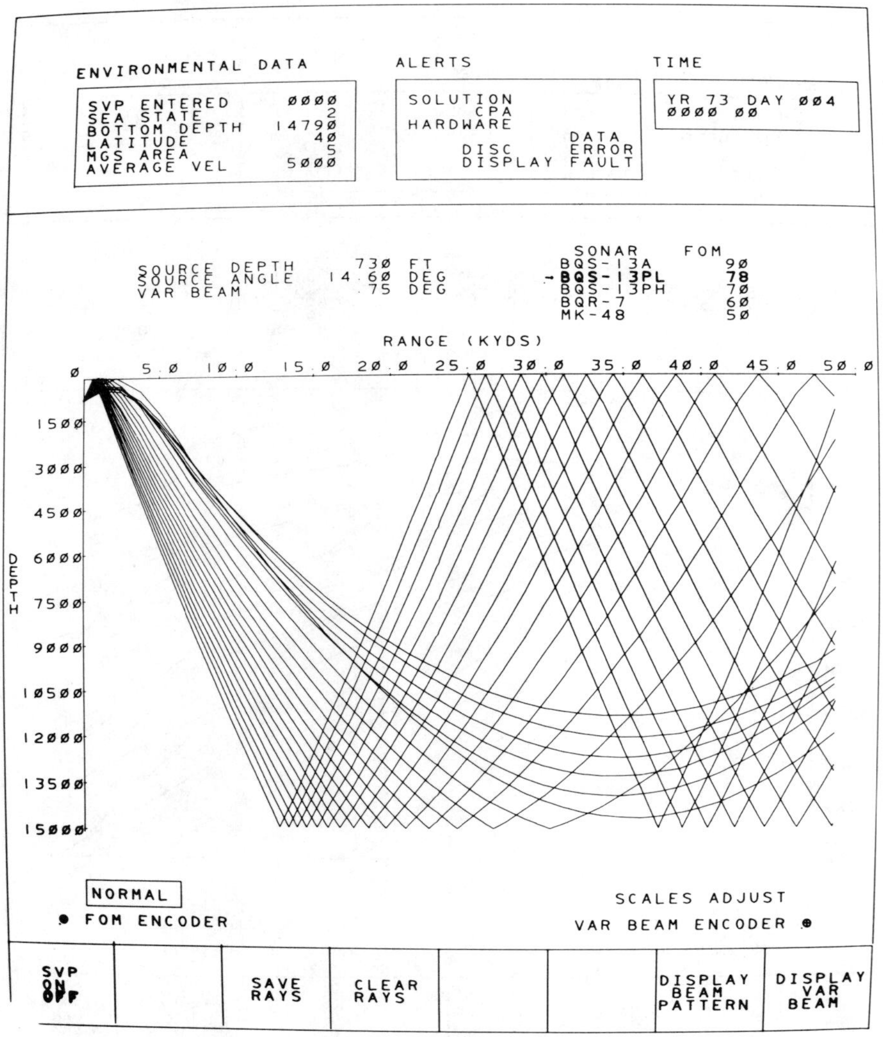

Source: Raytheon Company Photograph.

Figure 3-31. Ray Path Trace on the Raytheon MPS Display for Active Sonar Transmissions for a 12° Sonar Beam Directed Upward 14° Above Horizontal

for a sonar with a 12° beamwidth, with the beam directed upward at a 14° angle above the horizontal in the environment depicted in figure 3-25. Note that both a good convergence zone sound path and bottom bounce sound path exist together.

With a tool such as that discussed briefly above, the tactical commander is able to visually examine the predicted performance of his sonar. By varying the sonar parameters such as beamwidth or depression angle in the machine, he can select the values best suited for detecting submarines in the existing environment. On the submarine side, such a tool permits selection of best depth for search or evasion, as may be required.[b]

Notes

1. National Defense Research Committee (NDRC), "Principles and Applications of Underwater Sound," U.S. Government Printing Office, 1946, p. 2, para 1.2.2.

NOTE: The sonar equations as presented in this chapter are referenced to a 1 Hz bandwidth. Implicit in this is the assumption that both the sonar signal received and the sonar system receiving bandwidths are each 1 Hz wide, frequency-wise. When using these equations for predicting or comparing the performance of specific sonars, a corrective term must be applied to both the signal and noise terms of the sonar equation to reflect the actual or assumed signal and noise bandwidths, respectively. The corrective term in decibels is computed

$$10 \log W,$$

where W is the effective bandwidth concerned in Hz. When W is greater than 1, the computed corrective term is additive to the signal or noise term of the sonar equation as appropriate. When W is less than 1, the corrective term is subtractive.

The data for computing these corrective terms should be ex-

[b]Figures 3-24 through 3-31 were provided courtesy of the Raytheon Company, Submarine Signal Division, of Portsmouth, Rhode Island. These figures are photographs of the scope of the Raytheon Multi-Purpose Subsystem (MPS) now under test and evaluation at sea by the U.S. Navy.

pected to be found in the technical manuals accompanying the sonar. In extracting data from the sonar manuals, one should bear in mind that sonars will have different figures of merit and different bandwidths for different modes of operation.

4 Sonar and the Anti-Submarine Weapon System

In the preceding chapters, the emphasis and focus has been on the effects of the ocean environment on sonar performance, and on the use of the sonar equations and sound propagation loss curves to predict sonar detection performance. At this time, it is appropriate to refocus the discussion on the sonar in the perspective of anti-submarine warfare and as a component system or subsystem of the platform (surface ship, submarine, or aircraft) anti-submarine weapon system. In establishing this perspective, it is useful first to review anti-submarine warfare from an historical point of view over the last fifty or so years of war and peace.

Major Developments in Anti-Submarine Warfare from 1916 to 1974

During this period, the advantage has seesawed back and forth between submarine and anti-submarine units as a function either of total numbers of operational units, technological development, operational and tactical skills, or from a combination of two or more of these attributes. In World War I and during the first part of World War II, the submarine was essentially a vessel which transited and recharged its batteries running on the surface, and which usually submerged to conduct the attack. Submerged speeds were low (3-6 kts), torpedoes were straight running and preferably were fired from relatively short range (one half mile or less), and submarine operating depths were limited to about 350 ft. Surface anti-submarine units conducted sonar search and attack at 10-15 kts, using patterns of essentially zero-range weapons; depth charges dropped over or projected out from the stern and which exploded at preset depths, or short-range ahead-thrown bombs which required contact with the target to explode. Sonar ranges were limited to the direct path and seldom exceeded 1500-2000 yds. With its pronounced speed advantage over the submerged submarine, to attack, the destroyer depended on attaining a position over the submarine, and thence by launching successive patterns of the weapons described above, to obtain a hit and a kill. Since the submarine had to approach its target to close range, a screen of destroyers posed a barrier through which the submarine had to pass. In addition, in the early days of World War II, aircraft were valuable in searching for and attacking submarines on the surface or at periscope depth, using visual search and depth bombs. The later development and addition of radar to the aircraft increased its effectiveness manyfold and contributed greatly to eventual victory in the battle of the Atlantic in World War II.

Later in that war, the German development of the snorkel, which permitted the submarine to recharge its batteries while submerged and exposing only a short mast above the surface, provided a short-lived advantage for the submarine until higher resolution airborne radars and appropriate counter tactics were developed.

Toward the end of the war, the German type XXI submarine, with its higher submerged speeds (12-15 kts) and greater submerged endurance was a large step in the direction of a true submarine capable of completely submerged operations, but it came in too few numbers and too late to influence the outcome of the war.

Aside from the introduction of pattern running submarine torpedoes, the weaponry of both sides remained essentially as described above throughout the war, and anti-submarine tactics remained about the same.

The development and introduction of the nuclear powered submarines commencing in 1953 climaxed the post-World War II period of significant anti-submarine and pro-submarine development. Prior to the initial introduction of the nuclear powered submarine, both the United States and the Soviet Union had been building in quantity fast conventional submarines based on the German type XXI design, and submarine launched acoustic homing torpedoes were introduced. Concurrently, on the anti-submarine side, surface ship active sonars were much improved in both search rate and in figure of merit, and variable depth towed sonars were introduced to permit better detection capabilities against submarines below the thermal layer. Surface ship weapon ranges were extended out to 1000 yds. with various type mortar bombs, and anti-submarine acoustic homing torpedoes for both ship and airborne use were introduced to counter the high escape speed of the new high speed conventional submarines. Aircraft acoustic sensors, sonobuoys, were developed and introduced, and fast killer-type submarines were introduced into the anti-submarine forces.

The nuclear powered submarine rendered obsolete much of the hardware previously developed, and the traditional surface ship tactics as well. Specifically, the speed advantage of the destroyer over the submarine was eliminated as was the attack speed advantage of the then newly developed ASW torpedoes, and the conventionally powered killer submarines were greatly reduced in effectiveness. Even with subsequent improvements in surface ship active sonar and in longer range ASW weapons such as the U.S. ASROC (*A*nti-*S*ubmarine *R*ocket) or the Australian IKARA anti-submarine missile, and the high speed MK 46 torpedo, because of its speed, the nuclear powered submarine remains a very difficult target for a destroyer to bring under effective attack. In this connection, the addition of the manned anti-submarine helicopter as a weapon launcher to the destroyer weapon system has been essential to keep the destroyer effective in the anti-submarine business.

Since the introduction of the nuclear powered submarine, the most significant pro-submarine technological developments are the addition of the submerged launched strategic long-range missile, the surface launched cruise missile, the submerged launched multi-environment tactical missile, and increases in submarine speed up to or exceeding 30 kts. From the anti-submarine point of view, the significance of these four developments are several. The first and last two of the four signify the arrival of the true submersible, whose only need to surface, theoretically, is to replenish supplies and men. Thus, the usefulness of non-acoustic search sensors is greatly reduced. Second, they presage two primary target-oriented anti-submarine missions: the one directed against the patrolling strategic missile equipped submarines; the other directed against the anti-shipping missile/torpedo equipped submarine. Both of these type submarine groups are predominantly nuclear powered. Third, with respect to the protection of forces and units at sea against the long-range cruise missile equipped submarine, the total area around these forces required to be placed under acoustic surveillance or screening is vastly increased over that previously required to screen high value targets against the relatively shorter range submarine anti-shipping torpedo. This change is on the order of a circular area whose radius is 200 miles or so compared with a similar area whose radius was about 5 miles—125,000 square miles to 79 square miles. And last but not least, the increase in speed of the submarine to a position of superiority over many anti-submarine surface and submarine units requires reexamination of anti-submarine attack tactics and weapon capabilities.

It is conceivable that one could consider that the expected greater lethal destruction range of nuclear warheads on ASW weapons could neutralize the escape capability of the fast submarine. Certainly the authorization for the Navy to use such weapons would represent a political decision at the highest and most serious level. Perhaps an even more serious political decision, if made, would be the one to elect to have no alternative to using nuclear warheads in event of war at sea.

In any event, it is not likely that the results of any assessments of the current viability of the attack doctrine nor of the adequacy of their weapons made by any nation will be disclosed outside their highest government levels, so that any further discussion here would be well beyond the purpose of establishing an anti-submarine warfare perspective for sonar.

The nuclear powered submarine did not arrive on the scene without some liabilities which assist the anti-submarine forces. Its detectability by passive sonars is increased because its machinery plant radiates more noise (L_S in the passive sonar equation) than does a conventional submarine operating on batteries. If the steam power plant of a nuclear powered submarine is similar or analogous to that of a destroyer, some of its machinery noise such as reduction gear noise would be speed dependent, while other machinery units run at constant speed, like the turbine generator which provides electricity for the vessel. While efforts to reduce machinery radiated noise have been and will assuredly continue to be made, the achievable extent of such reduction can only be surmised.

A second effect adverse to the performance of the high speed submarine is the reduction to its sonar figure of merit resulting from increased self-noise at the higher speeds. The impact of this in detection range can be assessed by assuming a figure of merit at 1 kHz of 90 dB at 10 kts at sea state 2, obtaining a corresponding range from figure 3-2, and then obtaining a second range for comparison using a figure of merit reduced by the differential in self-noise (L_N) between 10 and 25 kts from figure 3-18. Using these data indicates a reduction in detection range from about 28 kyd to 2 kyd. The reduction in range using the propagation loss curve of figure 3-12 is even more marked, from about 34 kyds to 1 kyd.

From this brief review of the recent past to date, several useful conclusions can be made concerning the future capabilities needed by sonars, ASW ships, and ASW tactics for countering the cruise missile nuclear powered submarine. Because of the larger area to be screened around convoys and high value military ships, the first conclusion is that systems capable of a much higher search rate than at present are needed. By search rate is meant the area of ocean searched per unit time. In general terms, search rate is the product of ship speed and sonar sweep width, where sonar sweep width is approximately two times the predicted detection range. The magnitude of increase in sweep rate indicated needed in the above review suggests that increases in both ship speed and sonar sweep width are necessary unless the number of screening units are to be increased by orders of magnitude.

Second, the increase in figure of merit (and detection range) afforded passive sonars by the increased radiated noise levels of the nuclear powered submarine particularly when viewed in the light of the degradation in active sonar performance experienced against a fast submarine target suggests that passive sonar for search and target detection is a viable choice. A relatively new type of passive sonar and its tactical use will be discussed later.

Last, with respect to ASW surface ship speed, an increase in attack speed to a level sufficiently high to permit the attacking surface vessel to attain a position on top, or nearly so, of the target would afford the opportunity for more effective weapon placement, particularly with the present weapon inventory. From a sonar point of view, this suggests a need for a sonar capable of effective operation at very high ship speeds. Of course the use of aircraft to deliver sonobuoys to locate the target with sufficient accuracy to permit subsequent weapon launch from the aircraft is a desirable alternative attack mode which sometimes may be available to the ASW tactical commander on the scene. It is not unlikely that the best combined assets of both the ASW ship and its support aircraft are necessary to successfully place a weapon on a high speed nuclear powered submarine. The aircraft brings the necessary speed superiority to the engagement.

Evolving Sonar Screening Concepts

In the preceding chapters, an interdependency was identified between the ASW ship and its sonar in the context of sonar performance as evidenced in the elements of the sonar equation. In the discussions just above, a significant objective was an identification of the interdependency among the sonar, the ASW ship, and the weapon and its delivery system, in the execution of an anti-submarine mission. More specifically, the relationship between ship speed, target speed, and weapon placement was highlighted by the need for the use of aircraft by the ASW ship in order to achieve the requisite speed advantage over the submarine for effective weapon placement.

Even before the weapon placement phase of ASW engagement, however, is the search phase for the submarine. The magnitude of the screening problem was highlighted by the estimated area to be protected around a convoy or high value military target—a circle of radius about 200 miles or about 125,000 nautical miles. In the remainder of this chapter, the interaction of the tactical geometry of the anti-submarine encounter with the acoustic geometry of the ocean sound path structure and some additional design parameters of sonar systems will be examined. From a mathematical model of this interaction, screening concepts, including sonar and vessel types indicative of current trends and applicable to the magnitude of the screening problem, are discussed.

The perspective for this examination is the passive sonar, and more specifically the beam pattern and the system integration time of the passive sonar system.

Sonar Beam Patterns

Hull Mounted Sonar Beam Patterns

In the previous discussion of Directivity Index (NDI) (see p. 42), it was described in principle how beams were formed to achieve gains in the signal to noise ratio by noise rejection and directionality. Figure 3-5 in particular indicated how directional beams can be formed using various combinations of the hydrophone elements in the array. If the curved segment of the hydrophone array shown in figure 3-5 were extended to a closed circle representing the top view of a cylindrical array, and beams were formed all around the circle of elements, a 360° beam pattern would result, as is shown in figure 4-1.

Given such a pre-formed receiving beam, and knowing the fixed orientation of the elements in the array with respect to the ship's centerline, target bearing measurements can be made based on the direction of the beam axis (also called *m*ain *r*esponse *a*xis—MRA). In this situation, bearing accuracy is ½ beam width. Figures 4-2, 4-3, and 4-4 are pictorial representations of some typical beam patterns. Figure 4-2 is a typical beam pattern of a circular disc source, such as a single transducer element of an active sonar. Note that the beam pattern consists of a main lobe and numerous concentric side lobes, and that the main lobe extends out beyond the side lobes. The impact of side lobes on bearing error is apparent from the sketch, i.e., the wider the beam at the side lobe, the greater the bearing error (½ beam width). Suppression of side lobes is a desirable sonar design attribute, and is achieved electrically by a process called shading. (For a more detailed discussion of shading, see Urick.) The minus decibel (dB) notation in figure 4-2 is a way of indicating the beam pattern shape as a function of distance from the array. The tip or end of the main lobe corresponds to a signal to noise ratio of 0 dB—the distance from the array where the probability of detection of a given signal is 50%. The −10 dB point is at the distance from the array where the main lobe has its widest demension. At this distance from the array, the signal to noise ratio required to make a detection is 10 dB less than at the end of the main lobe. In figure 4-2, the first side lobe is at a distance where the signal to noise ratio is said to be 20 dB down (from the main lobe).

Figure 4-3 depicts the same beam pattern in both polar and vertical coordinates.

Figure 4-4 is the pictorial representation of a typical beam pattern of a cylindrical array—an array such as might result from combining a number of circular disc elements whose individual patterns are similar to that shown in figure 4-2. In this figure, the shaded beam width is indicated as the angle subtended by the −3 dB down points on the main lobe. By convention, unless otherwise indicated, when beam width is expressed in degrees, the beam width measured at the 3 dB down point (−3 dB) is understood.

Using the basic beam forming concepts shown in figure 3-5,

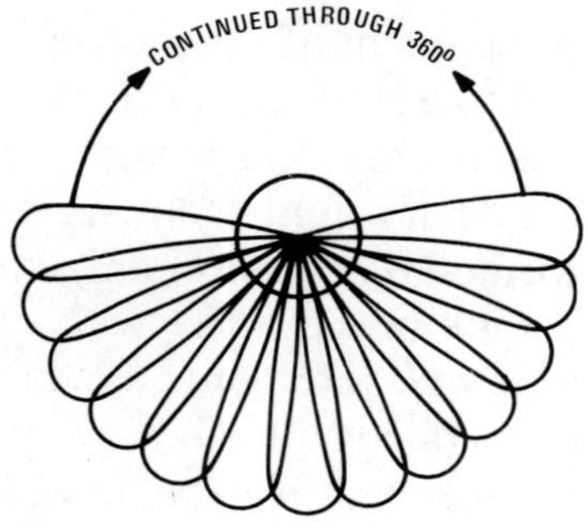

Figure 4-1. Typical Beam Pattern, Horizontal Plane, for Cylindrical Array

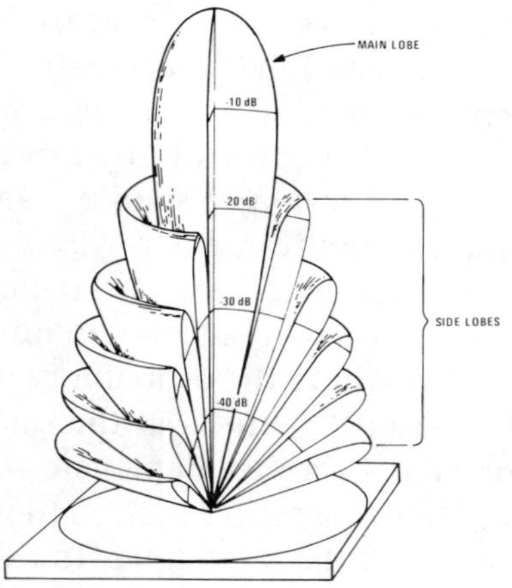

Source: *Introduction to Sonar Technology,* Bureau of Ships, Navy Department, Washington, D.C., 1965, fig. 4-2, p. 78.

Figure 4-2. Beam Pattern of a Circular Disk Source

beam formation or steering in the vertical plane as well as the horizontal may be done for cylindrical or vertical arrays. Figure 4-4 should therefore be viewed in the context as a single beam at a zero depression (or elevation) angle.

From the discussion above, it is seen that measuring target bearing from a hull mounted sonar array is relatively straightforward because the main response axis of each beam has a fixed orientation with respect to the centerline of the vessel. Given sufficient signal to

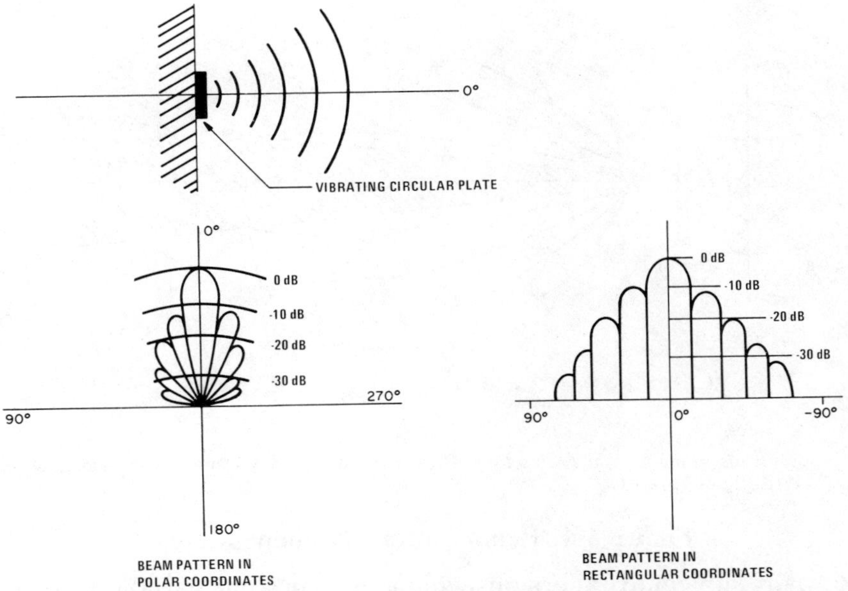

Source: *Introduction to Sonar Technology*, Bureau of Ships, Navy Department, Washington, D.C., 1965, fig. 4-4, p. 80.

Figure 4-3. Typical Beam Pattern for Circular Plate Transducer

noise ratio for a detection, target bearing can be measured directly. Bearing error is then a function of beamwidth. Some sonars have the capability of forming and steering to any direction a very narrow tracking beam to minimize bearing error. In addition, the component of bearing error attributable to roll and pitch of the vessel can be compensated to the horizontal plane using inputs from a gyroscope.

Towed Sonar Beam Patterns

Bearing measurements made from a towed variable depth transducer can be of comparable accuracy to those made with hull mounted transducers. The capability to do so is a function of designing sufficient hydrodynamic stability into the towed body as to maintain a fixed orientation of beam axis to ship centerline. A high degree of stability is achievable when the towing vessel is on a steady course. When the vessel is changing course, the orientation of the towed body axis with respect to the towing vessel centerline is

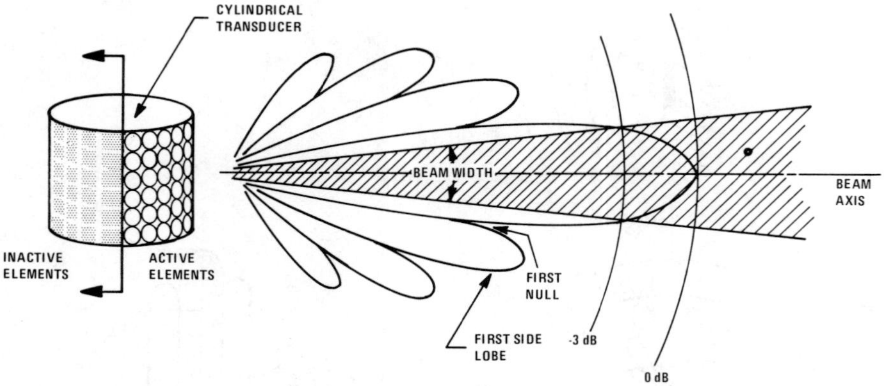

Source: *Introduction to Sonar Technology,* Bureau of Ships, Navy Department, Washington, D.C., 1965, fig. 4-5, p. 81.

Figure 4-4. Beam Pattern Nomenclature

normally subject to excursions which result in some variations in the sonar bearing error from that experienced when the vessel is on a steady course.

The need for greater acoustic detection ranges for screening vessels, generated by the development of the submarine launched long-range cruise missile, has resulted in renewed interest in towed linear array sonar systems. From points of view both of the propagation of sound in the ocean, and of the sonar equation, this is a logical development. First, sound propagation is attenuated less at lower frequencies (figures 2-6 and 3-12). Second, the radiated machinery noise levels of ships and submarines is greater at the lower frequencies (figures 3-22 and 3-23). Clearly, for passive sonars, the lower frequencies offer potential gains in detection range over the higher frequencies if practical sonar arrays to use them can be available.

Consider again the previous discussion of the task of the sonar designer to enhance the signal to noise ratio. This is done principally by the rejection of the noise background. This noise background is composed of both vessel generated noise and ocean ambient noise. From the sonar equation, the sonar directivity index (N_{DI}) is the design parameter principally associated with rejection of unwanted noise. An increase in N_{DI} would therefore enhance the signal to noise ratio and thus improve detection performance in range. From equation 3-10, for a simple line array, N_{DI} is calculated

$$N_{DI} = 10 \log 2L/\lambda$$

where L = array length and λ is the wavelength for the frequency concerned. Remembering that wavelength (λ) increases as frequency decreases, it is obvious that the effective array length must be increased significantly if significant gains in N_{DI} at lower frequencies are to be attained. For example, assuming a velocity of sound in water of 5000 ft./sec., the following are some nominal values of frequency and their associated wavelengths.

Velocity of sound–5000 ft./sec.	
Frequency (kHz)	*Wavelength (ft.)*
10.0	0.5
5.0	1.0
2.5	2.0
1.0	5.0
0.5	10.0
0.1	50.0

From this table, it is apparent that to achieve large gains in N_{DI} at lower frequencies, an array, physically very large, is needed. When considered in the context of the limited usable length available on the ASW hull for installation of such large arrays, the concept of a long towed linear array is very attractive. Additionally, if the acoustic receiving array is moved out of the immediate vicinity of the ship's own noise field, additional gains in noise reduction are possible. In this connection, note from figure 3-1 that 60 dB of propagation loss is normally experienced during the first 1000 yards of travel of the sound energy from its source due to spherical spreading loss.

A final aspect of the propagation of sound in the ocean, which increases the attractiveness of the passive towed acoustic linear array as an antisubmarine sensor, is the potential capability to tow it at depths where favorable sound propagation channels or ducts may be formed (see Chapter 2, p. 29).

While the beam pattern of the towed linear array has some commonalities with those of the cylindrical or spherical arrays described above, such as shape, side lobes, 3 dB and 10 dB down points, etc., it really differs significantly from that of the hull mounted sonars discussed above. The main lobe of each beam is a cone of revolution, and the total beam pattern is composed of a series of concentric cones of revolution about the array axis, as depicted in figure 4-5. The nature of this beam pattern presents some

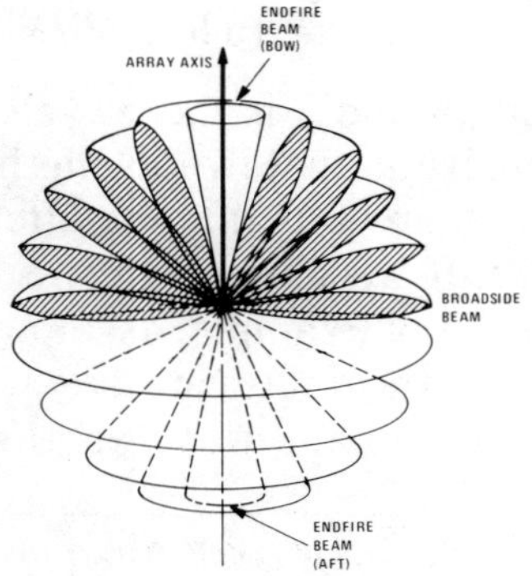

Figure 4-5. Typical Beam Pattern for a Towed Linear Array Sonar System

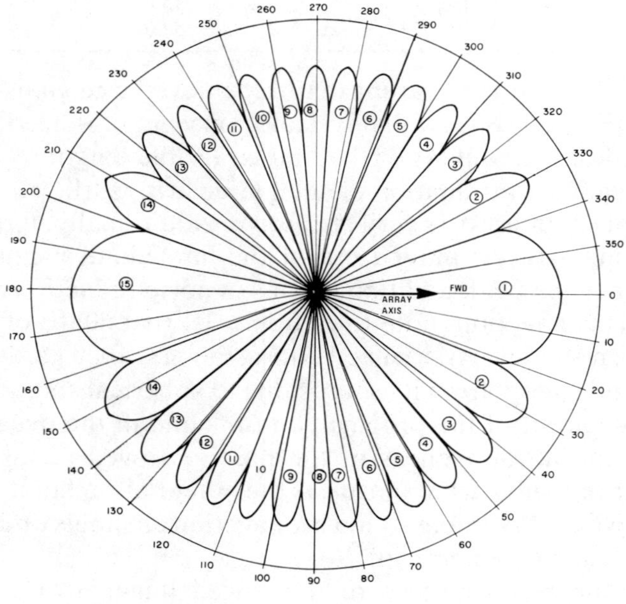

Figure 4-6. Horizontal Projection of the Beam Pattern of a 15 Beam Linear Array. [Circled numbers are beam numbers.] The lower half of the figure is the starboard side of the beam pattern. Note the port and starboard ambiguity for each beam.

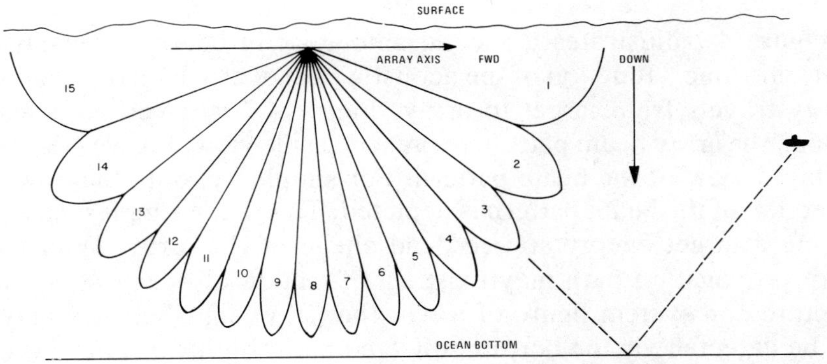

Figure 4-7. Side Projection of the Beam Pattern of a 15 Beam Towed Linear Array Showing the Bearing Ambiguity Resulting from a Target Signal Arriving via the Bottom Reflected Sound Path. In this case, a target nominally dead ahead is detected in beam No. 3, whose Main Response Axis is about 045° relative (Fig. 4-6.

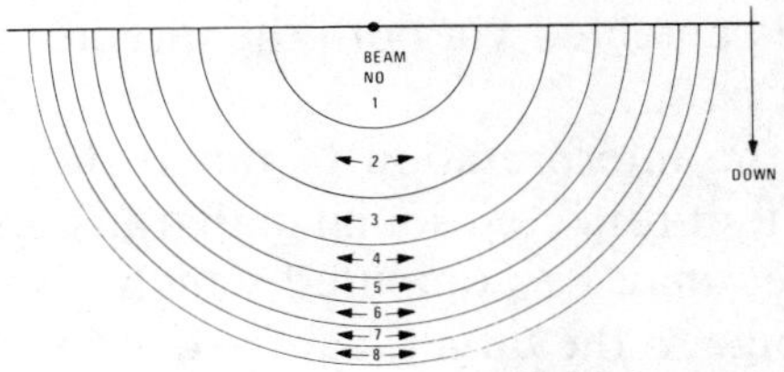

Figure 4-8. Front End View of the Beam Pattern of a Linear Array as Seen by a Target from Dead Ahead

complex problems in measurement of bearing. The nature of these problems is illustrated in figures 4-5, 4-7, and 4-8.

Figure 4-6 is the projection of the beam pattern on the horizontal plane—a horizontal slice through the beam pattern. Note that with the exception of the end-fire beams (beams 1 and 15), each beam has a port and starboard projection. In other words, when a target appears in any beam except the end-fire beam, there is a bearing ambiguity—the target may be either to starboard or to port. This ambiguity is readily resolvable by a simple turn maneuver of such magnitude as to ensure that the target will be re-detected in another beam. The magnitude of the apparent target movement of the false or ambiguous target readily resolves the ambiguity.

Figure 4-7 illustrates a second ambiguity in bearing measurement, this time a function of the acoustic path over which the sound energy travels from target to array. Figure 4-7 is a vertical slice through the array beam pattern to present a side view. Figure 4-8 is a front end view of the beam pattern. For simplicity, only the down projection of the beam pattern is depicted. To use the simplest case, assume a target energy source dead ahead of the array. While a direct propagation path may exist as a function of ocean velocity structure and system figure of merit, the target signal energy may also be detected via the bottom reflected path. In this instance, the target is shown detected in beam No. 3. From the horizontal projection of beam No. 3 (figure 4-6), the magnitude of the bearing error is the displacement of the MRA between beam No. 1 and beam No. 3. If the target were only detected in beam No. 3, a maneuver to place the target in the broadside beam would eliminate the apparent error of the bottom reflected path. The resolution of this bearing ambiguity concurrent with that of the port and starboard ambiguity may be significantly more complex than a single maneuver, particularly if the same target is detected via both the direct sound path and the bottom reflected path.

To assist in the interpretation of target data presented where multi-path or bottom-reflected sound path transmission is expected, a set of templates may be computed which, for the actual water depth and the range to the target, can be overlayed on the nominal horizontal beam pattern of the array. These overlays may be constructed using the following relationship:

$$\theta_I = \cos^{-1} \cos \theta_A \cos D/E \qquad (4.1)$$

where θ_I is the relative target bearing angle (measured from the bow) as indicated on the sonar display,

θ_A is the actual relative target bearing measured from the bow, and

D/E is the depression angle from the horizontal of the actual incoming sound signal.

It can be shown that

$$D/E = \tan^{-1} 2d/R \qquad (4.2)$$

where d is water depth in yards and R is actual range to the target (yds.). Substituting,

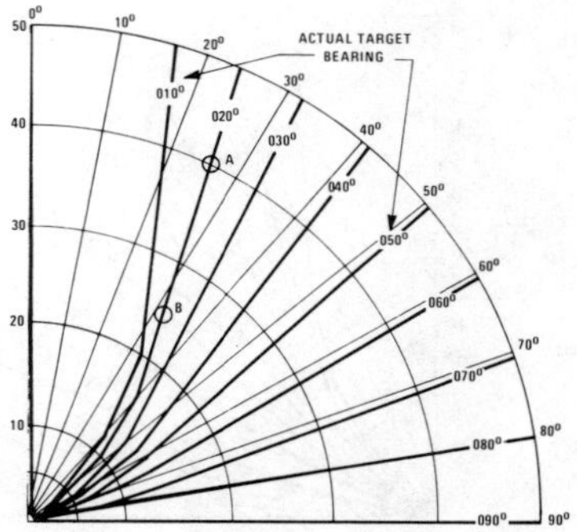

Figure 4-9. Plot of Curves of Indicated Target Bearing vs. Actual Target Bearing for a Water Depth of 2000 Fathoms over the Range Band 0-50 Miles for a Nominal 15 Beam Linear Array. Actual target bearings are as indicated. The indicated target bearing for a target at 40 miles whose actual bearing is 020° (pt. A) will be 026°. If the target closed to pt. B (020° 25 miles), the indicated target bearing as seen by the system would be 032°.

$$\theta_I = \cos^{-1}[\cos \theta_A \cos(\tan^{-1} 2d/R)]. \tag{4.3}$$

To calculate an overlay for a given depth, a water depth is assumed and for every ten degrees of relative bearing from 010° to 170°, a value of θ_I is calculated for ranges, for every 10 miles downward from a range such as 50 miles. On a polar plot to scale, values of θ_I for each assumed value of θ_A are plotted and connected with a smooth curve. The resultant plot should be similar to figure 4-9. If a template is made of this plot and overlaid on a beam pattern plot which is scaled to predicted bottom reflected range capability based on calculated FOM, the actual displacement of θ_I from θ_A is visible by inspection. Correlating this with changes in θ_I as observed on the display can be of assistance in resolving the ambiguity, or in determining if a sound path ambiguity does, in fact, exist. Figure 4-10 is an example of the plot from figure 4-9 overlaid on a nominal beam pattern plot.

Based on this discussion, one might be tempted to the conclusion that the potential of the towed array as a long-range passive acoustic sensor is seriously jeopardized because of the bearing ambiguities resulting from the nature of its nominal beam pattern and the interac-

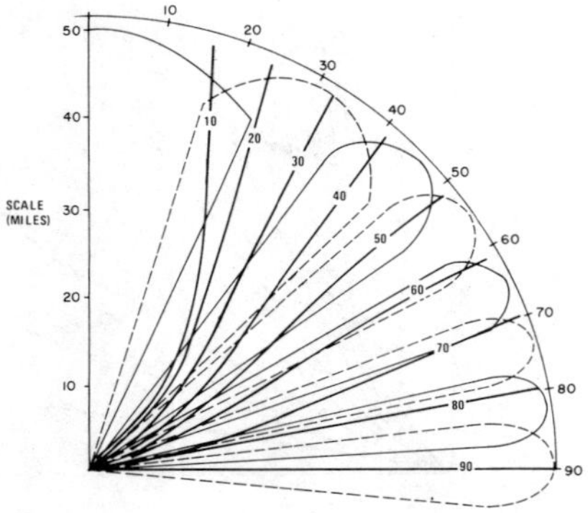

Figure 4-10. Overlay of the Curves of Figure 4-9 on One Quadrant of a Nominal 15 Beam Linear Array Beam Pattern

tion of this beam pattern with both the tactical geometry and the acoustic geometry of the antisubmarine encounter. Considering that the problem is susceptible to mathematical modeling, and given the computation capability in modern digital computers, which can be available on board ship, such a conclusion is not justified. The necessary tools to assist in resolving these problems in real time are capable of being brought into being with existing technology.

Sonobuoy Sonar System Beam Patterns

The sonobuoy sonar system is really a hydrophone array suspended from a buoy on the surface of the ocean "connected" by radio to processing and display equipment located on a ship or aircraft. Like hull mounted sonar systems, sonobuoy systems may be either active or passive. Passive acoustic signature data or target echoes for an active system are radioed back to the parent platform where sonar operators make detection and classification decisions. A significant difference between many sonobuoys which utilize only one hydrophone in the array, and other systems with multi-hydrophone arrays, is the difference in Directivity Index, N_{DI}. For a sonobuoy using a single omnidirectional hydrophone, N_{DI} is zero, and target bearing cannot be measured. Other sonobuoys in use by the U.S. Navy include directional or direction-measuring sonobuoys, for which values of Directivity Index of up to 5 dB are sometimes claimed.

Bearing measurement from a single hydrophone requires a

multi-element hydrophone oriented to a predetermined magnetic heading and four sensitive elements of the hydrophone mounted 90° apart from each other in the horizontal plane. Assume that the orientation of the hydrophone to the magnetic compass is such that two of the hydrophones line up in a North-South direction; then the other two lie along an East-West axis. Electrically connecting the N-S and E-W hydrophone element pairs respectively forms a N-S and an E-W dipole receiving beam pattern, as illustrated in figure 4-11.

Figure 4-12 illustrates the geometry of the incoming wave with respect to the N-S element pair separated by distance *d*. A plane wave incident to the N-S axis at angle ϕ is detected by element #2 t seconds after being detected by element #1. In right triangle abc, the length of leg ℓ is calculated $\ell = tc$, where t is the measured time difference in signal arrival at elements 1 and 2, and c is the velocity of sound in water. Angle θ, the bearing of the incident plane wave with respect to the N-S axis can be calculated on the basis that

$$\cos \phi = \ell/d; \quad \phi = \cos^{-1} tc/d.$$

Note that there is another triangle abc′ which also satisfies this relationship. It is apparent that with the dipole receiving pattern, there is a bearing ambiguity, with one candidate bearing in the NS quadrant, and another in the adjacent NE quadrant.

Similarly, for an E-W oriented element pair, it could be shown that the plane wave signal depicted in figure 4-12 would cause candidate bearings in the NE and SE quadrants respectively. To resolve the ambiguity, the signal levels for each quadrant, as computed above, are summed and compared with the omnidirectional signal level. If all hydrophone elements were equally sensitive, the summed signal level of the quadrant from which the signal emanates would be twice that of the signal level received at any one hydrophone.

It should be apparent that if this type sonobuoy is looking at two different sound sources in the same frequency region on different bearings, they will appear to the sonobuoy to be a single source at an intermediate bearing.

System Integration Time

Another key design parameter for passive sonars which interacts

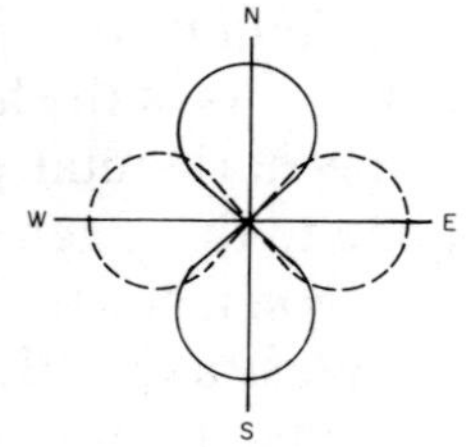

Figure 4-11. Typical Beam Pattern for a Four-Element Hydrophone

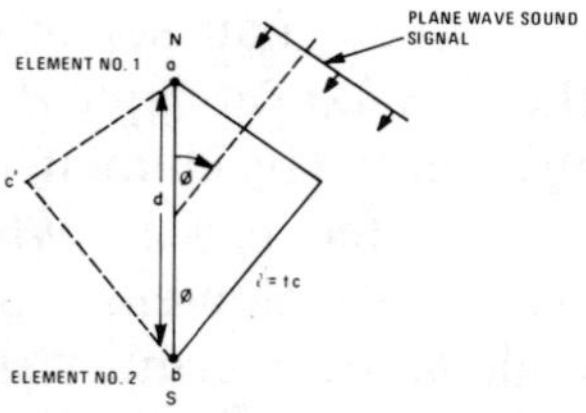

Figure 4-12. The Direction-Resolving Geometry of a Plane Wave of Sound Energy and a Two-Element Linear Array

with the tactical geometry of the anti-submarine encounter situation is system integration time. System integration time may be defined in a number of ways. First, it may be defined as the time the target energy field must be available to the system for a detection to be made. Being available to the system means being within the nominal predicted system detection envelope. Second, it may be defined as the duration of time required for the system to accumulate target signal energy measured from the first indication of the presence of a target energy field until sufficient signal has been accumulated to permit a detection by the sonar operator. And third, it may be defined as system detection time. Unlike the active sonar where a short sharp pulse of transmitted energy is reflected from the target as an echo of duration comparable to that of the transmitted pulse, passive sonars depend on a process involving the summation of target signal energy over a finite period of time to detect their presence.

The concept of system integration time is related to the manner in which the presence of target energy fields is presented to the passive sonar operator. While any one of a number of display formats could be used to illustrate the point, the frequency vs. time waterfall display will be used here. Consider a passive sonar system where each beam has a separate display, and that the nature of the display is signal frequency on the horizontal scale vs. time on the vertical scale. The output of the system signal processor is transmitted to an electrographic recorder or (cathode ray tube (CRT) display). The pen of the electrographic recorder (or the sweep of the CRT) passes across the paper on the recorder (or the screen of the CRT) from left to right at a fixed predetermined rate of speed. On the electrographic recorder, the paper advances downward so that a time history of successive sweeps is presented to the operator. A similar time history is presented electrically on the CRT display. If a target signal is present at some frequency (f), the intensity of the recorder pen trace (or the electronic sweep) at that frequency is increased according to the signal to noise ratio. Bearing in mind that an average signal to noise ratio of zero is the criteria for a detection decision, and remembering the standard deviation of N_W on the order of 5-6 dB, it is apparent that on some sweeps of the pen (or electronic sweep), the presence of an energy field will be indicated, and on other sweeps it will not. Given the presence of a target energy field just entering the sonar performance envelope, a number of sweeps will be necessary before an affirmative detection decision can be made. The time required to produce this requisite number of sweeps is the system integration time. In the sonar system design process, the resolution of the signal processor, the resolution characteristics of the display, the sweep rate of the display, and the pre-established requirements for probability of detection and probability of false alarm are utilized in establishing system integration time.

Figure 4-13 is a simplified sketch of a typical frequency vs. time waterfall display.

Availability of the Target to the Passive Sonar System

For passive sonar systems to detect a target energy field, the target must be "available" to the sonar in both time and space. From a spatial point of view, target availability is normally considered in the context of the target being inside the sonar's FOM detection range corresponding to the propagation loss for the environment concerned. However, target availability has a second important dimension in range; that of the minimum range at which the target is

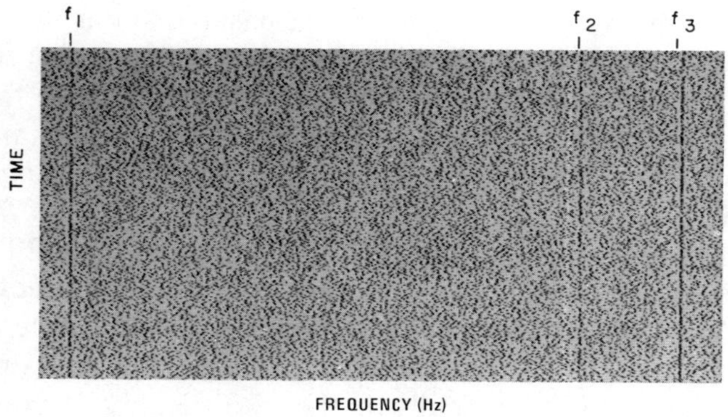

Figure 4-13. Frequency vs. Time Waterfall Display Format, Showing Signals of Frequencies f_1, f_2, and f_3. The concept of detection time is illustrated by viewing the display in successively increasing time window intervals until each target signal is discernible from the background noise. An initial time window of about 1/16 inch is suggested.

sufficiently available to the system in time as a function of sonar system parameters interacting with the tactical geometry of the anti-submarine encounter. Figure 4-14 depicts such an encounter where ship S encounters a target T.

Target T's relative motion vector with respect to ship S is traversing one beam of the ship's passive sonar along the line $a-a'$. Track $a-a'$ intersects the main response axis (MRA) of the sonar beam at I, making an angle φ with the MRA. The distance from ship S to I is range R. Sonar beamwidth is θ degrees. Let R be the minimum range at which the target can cross the sonar beamwidth θ on track $a-a'$ and still have the target in the beam a sufficient time to make a detection. It can be shown that

$$R = \frac{2V_R t}{\theta \cos \theta/4} \times \frac{\sin (\varphi - \theta/2) \sin (\varphi + \theta/2)}{\sin (\varphi + \theta/2) + \sin (\varphi - \theta/2)} \tag{4.4}$$

where R is the minimum range at which detection can take place in the sonar beam as a function of θ, t, φ, and V_R,

θ is the beamwidth,

t is the system integration time,

φ is the angle of target relative motion vector to sonar beam mean response axis, and

V_R is the relative speed of traversing the sonar beam by the target.

From this relationship, it is readily apparent that the minimum detection range (R) increases with (a) an increase in system integration time (t), (b) an increase in relative target speed with respect to the array (V_R), (c) a decrease in system beamwidth (θ), and (d) an increase in angle of crossing the MRA (φ).

The minimum range at which the target is available is of interest for several reasons. First is to determine if the target is available to the system at all, given the minimum range calculated above and the predicted detection range based on FOM and the appropriate propagation loss curve. If the minimum range exceeds the FOM prediction, the target is not available to the system without some change in the system parameters or in target relative motion. Second is to determine that if the target is available, at what range will it become not available to the system as the range is closed. Third is to determine if the passive search sonar is able to hold contact down to a range where the attack sonar can obtain contact; in other words can a sonar-to-sonar handoff of the target be accomplished without losing contact. The final reason is the capability of the high speed

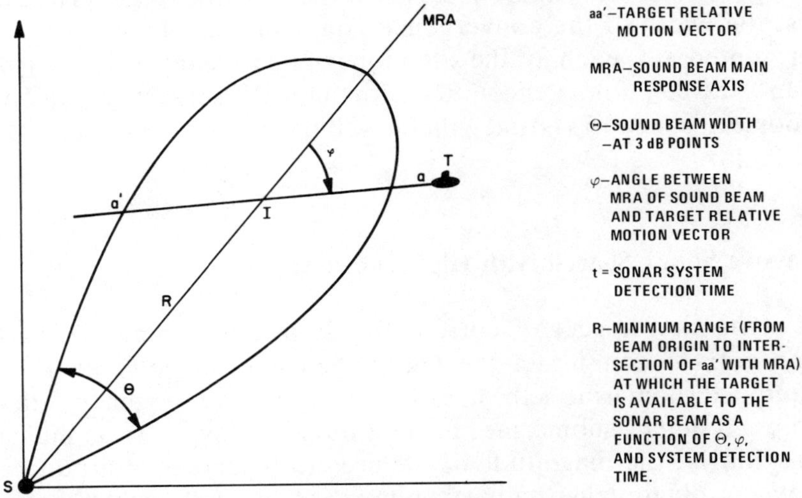

Figure 4-14. Geometry of the Intersection of the Target Relative Motion Vector with a Passive Preformed Receiving Beam

submarine to reduce its availability, and thus its vulnerability to attack, by running at high speed in any given transit may also reduce its vulnerability to detection and tracking by a given sonar notwithstanding the higher radiated noise level generated by the high speed.

Table 4-1 is a set of tables of values of R calculated for some random nominal values of θ and t, and for a range of values of V_R and φ that encompass the tactical geometries anticipated. Inspection of the tables illustrates the variability of R as a function of the system parameters θ and t, for a series of tactical encounter geometries, as well as the magnitude of the ranges that are involved. The value to the operator of a capability to vary the system integration time to reduce the minimum detection range against a high speed target approaching close-in is also illustrated.

There is one additional interaction between the tactical geometry of the anti-submarine encounter and the acoustic geometry of the sea of tactical significance that affects the availability of the target to the sonar system. That interaction is the relative speed of penetration of the target across the convergence zone annulus. The variables in the tactical geometry are own ship speed, target speed, and relative target bearing. The variable in the acoustic geometry is width of the convergence zone annulus. The magnitude of this interaction on the availability of the target to the system is illustrated in figure 4-15. This figure is a plot of the time the target is available vs. the width of the convergence zone annulus. Three curves are presented, one each for the situations of relative target bearing 000°, 045°, and 090° for an encounter with the ASW searching ship on one course at speed 15 kts and with the submarine on an intercept course at 25 kts.

Passive Sonar Search with High Speed Ships

A significant aspect of current trends in anti-submarine warfare concepts to counteract the high speed nuclear-powered missile-equipped submarine is the use of passive acoustic sensors by surface ships to detect submarines at long ranges. However, as indicated previously, the magnitude of the area to be screened around convoys and high value military shipping at sea will probably require both longer range sonars and high speed platforms to provide effective coverage of the area with reasonable number of ships. Two emerging high speed surface ship types appear admirably suited to

Table 4-1
Table of Minimum Detectable Range R (kyds) as a Function of System Integration Time t, Beamwidth θ, Relative Target Speed V_R (kts), and Target Angle Crossing the Sonar MRA

A. $\theta = 16°$, $t = 600$ sec.
$\theta = 8°$, $t = 300$ sec.
$\theta = 4°$, $t = 150$ sec.

V_R	15°	30°	60°	90°
8	2.5	4.7	8.1	9.5
12	3.7	7.0	12.3	14.2
18	5.6	10.6	18.4	21.3
24	7.4	14.1	24.5	28.5
27	8.3	15.9	27.7	32.0
36	11.2	21.1	36.8	42.6
48	14.8	28.2	49.0	57.0

B. $\theta = 16°$, $t = 300$ sec.
$\theta = 8°$, $t = 150$ sec.
$\theta = 4°$, $t = 75$ sec.

V_R	15°	30°	60°	90°
8	1.3	2.4	4.0	4.7
12	1.9	3.5	6.1	7.1
18	2.8	5.3	9.2	10.7
24	3.7	7.0	12.3	14.3
27	4.2	7.9	13.8	16.9
36	5.7	10.5	18.3	21.3
48	7.4	14.0	24.5	28.6

C. $\theta = 16°$, $t = 150.0$ sec.
$\theta = 8°$, $t = 75.0$ sec.
$\theta = 4°$, $t = 37.5$ sec.

V_R	15°	30°	60°	90°
8	0.7	1.2	2.0	2.4
12	1.0	1.8	3.0	3.6
18	1.4	2.6	4.6	5.4
24	1.9	3.5	6.1	7.1
27	2.1	4.0	6.9	8.0
36	2.8	5.3	9.1	10.7
48	3.7	7.0	12.2	14.3

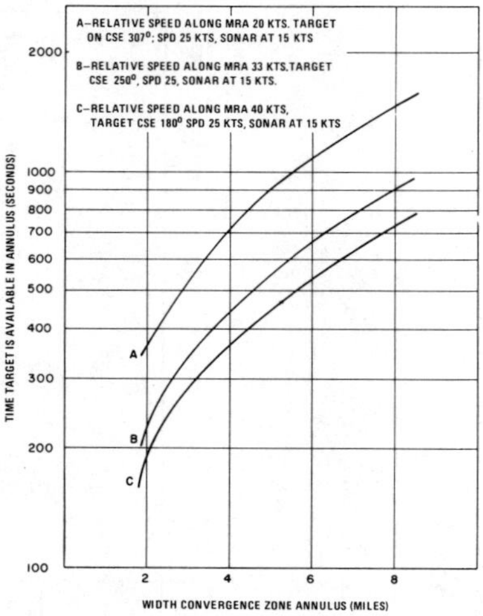

Figure 4-15. Plot of the Time the Target is Available to the Sonar System as a Function of CZ Annulus Width for Three Representative Encounter Geometries

this task, the hydrofoil and the surface-effect ship (air cushion vehicle). Recent literature indicates projections of 50 kt, 350-500 ton hydrofoils, and 90 kt surface effect ships of displacements up to 2000 tons. Ship speeds of this magnitude not only contribute greatly to increased sonar search rate, but also provide a speed advantage over the submarine adequate for effective weapon placement.

The suggested use of high speed ships as sonar platforms may appear to be at odds with sonar performance as manifested by the sonar equation and the concept of sonar figure of merit. And, as figure 3-18 would indicate, to attempt to conduct passive sonar search with a towed linear array at the speeds projected would be futile, to say the least. There is, however, a concept for utilizing both the long range detection capability of the towed linear array sonar, and the high search speed of the hydrofoil on the surface-effect ship. This concept, called the *sprint and drift* concept, provides for conducting passive acoustic search at acceptable sonar speeds for a predetermined length of time, and then sprinting at high speed to a new location, where speed is again slowed to acceptable sonar search speed, and the cycle is repeated. The implementation of this concept for area search involves a detection range prediction based on figure of merit, the sonar system detection (integration) time, the time interval required for the towed linear array to stabilize in performance after the sprint, and an assumption as to target speed.

For convoy screening, an additional factor introduced is the relative speed of the screening vessel with respect to the speed of the screened force. Implementations of this concept are discussed in detail below.

First, a word concerning stabilization of towed array performance. From previous discussions of the spatial relationship between hydrophone elements in forming pre-formed receiving beams and the use of these beams to measure target bearing, it should be apparent that the orientation of the array axis to the direction of the towing ship centerline is critical. Likewise, from the discussion of the sound velocity structure of the ocean and the resulting sound path structure, it is apparent that best performance will be attained with the array at some particular depth at any given time. A course change by the towing vessel disturbs the axial orientation of the towed linear array, a speed change results in a change in array depth. Therefore, after each such maneuver, a certain period of time must elapse before the array is sufficiently stable to achieve design performance. This time is commonly called array stabilization time. It will vary with different arrays as well as with changes in course and/or speed.

Second, before proceeding with a detailed discussion of sprint and drift tactics, the concept of probability of detection as a function of availability of the target to the sonar system will be discussed. In previous discussions of target availability, the focus was on the determination of the availability of the target for one detection (integration) time period as a function of the tactical geometry and the acoustic geometry of the encounter situation. In this context, the assumed probability of detection for one detection time period was 50%, according to the concepts of the sonar equation. If the target is available to the system for more than one detection period (one look), the cumulative probability of detection is increased, according to the number of independent "looks" available. If each successive system integration time period is considered a separate independent look into the same body of water, or at the same target, then the probability of detection may be calculated according to the following:

$$P\text{cum} = 1 - (1 - P_1)^n \tag{4.5}$$

where Pcum is the cumulative probability of detection,

P_1 is the probability of detection for a single observation, and

n is the number of observations made.

Some nominal values calculated from this relationship are as indicated below

	No. of observation (n)				
	1	2	3	4	5
Pcum (%)	50	75	87.5	93.5	97

There is yet a third additional concept involved in sprint and drift search and/or screening, that of the size and shape of the area maintained under surveillance by the acoustic sensor. The traditional approach is to view the sonar performance envelope as a circle whose radius (R) corresponds to the 50% probability detection range computed according to the allowable propagation loss in the environment specified for system figure of merit, as in figure 4-16. If a target traverses the array performance envelope along the track OAB, then the target is available to the system for the time AB/V_R, where V_R is the relative speed of the target with respect to the array along the track AB. It can be seen that AB, and thus the time the target is available to the system, varies through a range of values from 0 to $2R$. For practical analytical purposes, a smaller range of values expressed in terms of R is desirable. Further, because targets may pass through the effective performance envelope of the sonar from any direction, a sonar performance envelope shaped as a square oriented to the direction of relative movement of the target through the envelope is suggested. Figure 4-17 shows two different such squares superimposed on the circle of radius R.

In this convention, the front-to-back distance of the sonar is called *E*ffective *S*creen *D*epth (ESD). In figure 4-17, square ABCD is formed by connecting points 90° apart on the circumference of the circle. The length of one side or Effective Screen Depth (ESD) is $1.42R$. Square EFGH is the square whose area is the same as that of the circle with radius R. ESD for this square is $1.76R$. Because of the similarity of this value to that commonly used in close-in screen ship spacing, a square whose side (ESD) is $1.75R$ (predicted range for 50% probability of detection) is the writer's choice, and will be used hereinafter unless specified otherwise.

Consider now the encounter situation depicted in fiqure 4-18.

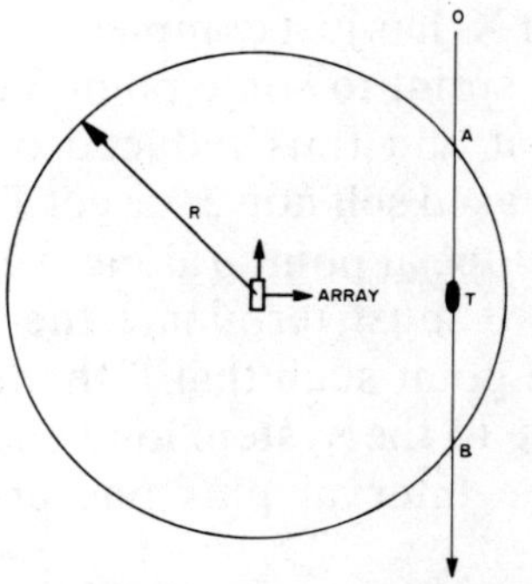

Figure 4-16. Target *T* Traversing the Detection Performance Envelope along the Line OAB of Sonar with Detection Capability *R*

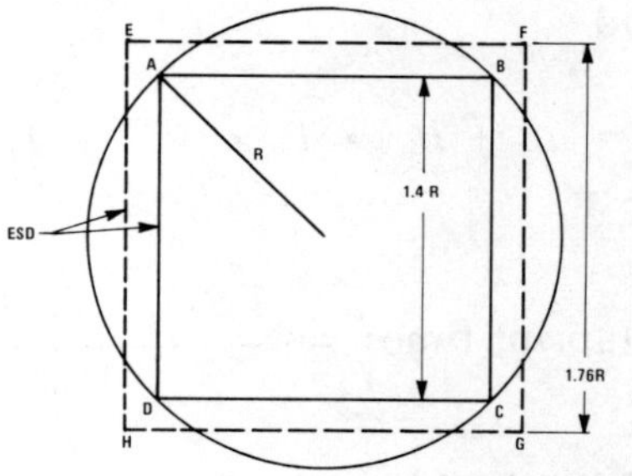

Figure 4-17. Concept for Effective Screen Depth (ESD) based on Sonar Detection Performance Envelope Radius *R*

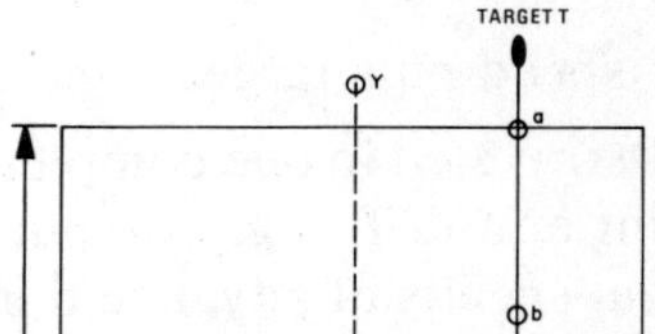

Figure 4-18. Sprint and Drift Search Concept

Screen ship S, at point X, has just completed one look with his sonar, and is commencing a sprint to some point Y. Concurrently, just as ship S began to sprint and thus reduced or eliminated his sonar capability due to increased self noise, target T began a penetration of S's performance envelope at point *a* along the line abc. If ship S is to detect target T, then it must terminate the sprint and take sonar search speed at some point such that if the target is at point *b*, the target will be available to the system long enough for one look (one system detection time interval plus one array stabilization time interval).

Let T_S be sprint time, T_L be system time, V_{RS} the relative speed of penetration of the target with respect to the array during sprint, and V_{RL} the relative speed of target transit through the performance envelope during system time. Then, from figure 4-18, $ab = T_S \times V_{RS}$, $bc = T_L \times V_{RL}$, and

$$\text{ESD} = ab + bc = T_S \times V_{RS} + T_L \times V_{RL} \tag{4.6}$$

Transposing,

$$T_S \text{ (sprint time)} = \frac{\text{ESD} - T_L \times V_{RL}}{V_{RS}}, \tag{4.7}$$

where ESD = 1.75R (FOM prediction)

T_L = System detection time plus array stabilization time,

V_{RL} = sonar search speed plus target speed, and

V_{RS} = sprint speed plus target speed.

The actual distance (D) traveled in one complete cycle is the sum of travel during the sprint and drift leg, computed using T_L and T_S, respectively with actual speeds of advance during each phase. Let V_S be sprint speed and V_L be sonar search speed, then

$$D = T_L \times V_L + T_S \times V_S \tag{4.8}$$

Speed made good (V_G) is calculated

$$V_G = \frac{D}{T_L + T_S} = \frac{T_L \times V_L + T_S \times V_S}{T_L + T_S} \tag{4.9}$$

Search rate (SR) is defined as search width (ESD) times average speed (V_G) and is calculated.

$$SR = ESD \times VG = 1.75R \frac{T_L \times V_L + T_S \times V_S}{T_L + T_S} \quad (4.10)$$

Search rate is commonly computed in square nautical miles/hr. in the United States, and in square kilometers/hr. in the metric system.

It can be seen that maximum search rate occurs when the probability of detection needed is only 50%, that is, when T_L is computed on the basis of one system detection time.

For illustrative purposes, assume the following:

Target speed 25 kts
Sonar search speed 15 kts
Sprint speed 50 kts
Sonar detection range prediction 50 miles
Sonar detection time 10 min.
Array stabilization time 20 min.

Solve for: Speed made good during one complete cycle, and the search rate achieved for a 50% probability of detection.

Solution $V_L = 15\text{kts}$ $T_L = 0.5 \text{ hr}$
$V_S = 50 \text{ kts}$ $R = 50 \text{ miles}$
$V_{RL} = 40 \text{ kts}$ $V_{RS} = 75 \text{ kts}$

$$T_S = \frac{1.75 \times 50 - 0.5 \times 40}{75} = .09 \text{ hr or } 54 \text{ min.}$$

$$V_G = \frac{0.50 \times 15 + 0.9 \times 50}{0.9 + 0.5} = 37.5 \text{ kts.} \quad \text{ANS.}$$

$$SR = 87.5 \times 37.5 = 3280 \text{ n. mi.}^2/\text{hr.} \quad \text{ANS}$$

Next, solve for the speed made good during one complete cycle and the search rate for the following cumulative probabilities of detection, respectively.

75%, 87.5%, 93.5%, and 97%.

Solution: First, compute T_L for each of the above situations according to the number of detection time intervals required to achieve the desired probability of detection.

For 75% P/D, T_L = 2 × 10 + 20 min., or 40 min. (0.66 hr.)
87.5% P/D, T_L = 3 × 10 + 20 min., or 50 min. (0.84 hr.)
93.5% P/D, T_L = 4 × 10 + 20 min., or 60 min. (1 hr.)
97.5% P/D, T_L = 5 × 10 + 20 min., or 70 min. (1.16 hr.)

Next calculate sprint times for each of the above drift times.

For 75% P/D, T_S = 0.815 hr., or 49 min.
87.5% P/D, T_S = 0.720 hr., or 43 min.
93.5% P/D, T_S = 0.635 hr., or 38 min.
97.5% P/D, T_S = 0.585 hr., or 35 min.

Next calculate V_G

For 75% P/D, V_G = 34.5 kts.
87.5% P/D, V_G = 31.5 kts.
93.5% P/D, V_G = 28.5 kts.
97.5% P/D, V_G = 26.5 kts.

Finally, calculate the respective search rates (SR)

For 75% P/D, SR = 2980 n. mi.^{2}/hr.
87.5% P/D, SR = 2730 n. mi.^{2}/hr.
93.5% P/D, SR = 2470 n. mi.^{2}/hr.
97.5% P/D, SR = 2290 n. mi.^{2}/hr.

It is useful to group the above calculations in tabular form as follows.

Cumulative Probability of Detection (%)	*Drift Time (hrs.)*	*Sprint Time (hrs.)*	*Speed Made Good (kts)*	*Search Rate (n. mi²/hr)*
50.0	0.50	0.90	37.5	3280
75.0	0.66	0.82	34.5	2980
87.5	0.84	0.72	31.5	2730
93.5	1.00	0.64	28.5	2470
97.5	1.16	0.59	26.5	2290

From inspection of this table, for the predicted detection range, target speed of penetration and vessel operating speeds assumed, the following conclusions are made for the hydrofoil and towed array sonar system combination:

1. There is a capability for a detection probability of 50% while screening mobile forces whose speed of advance is 36.5 kts;
2. An 87.5% probability of detection can be achieved while screening forces whose speed of advance is equal to or less than 31.5 kts.
3. A 97.5% probability of detection can be achieved while screening forces whose speed of advance is equal to or less than 26.5 kts.
4. For speeds of advance of less than 26.5 kts, lower hydrofoil operating speeds could be used to achieve comparable capabilities to those cited in 1, 2, and 3 above. For example, a 97.5% probability of detection can be achieved with a search speed of 10 kts and sprint speed of 35 kts while screening a force whose speed of advance is 20 kts.

If the target is penetrating at lower speeds than that assumed, then the target is available for a longer period of time during the sprint and drift cycle, and sonar detection probability would be improved. A reduction in predicted detection performance in range, on the other hand, would reduce the availability of the target, and sonar detection probability would be worsened.

In the examples above, only the hydrofoil sprint speed range has been considered, but both slower and faster speed platforms can use sprint and drift screen (or search) tactics effectively. At the lower speed end, consider a screened force with SOA (speed of advance) of 20 kts; screened by conventional surface escorts utilizing towed linear array sonar systems. Search speed is 15 kts, sprint speed 25 kts, T_L is 0.5 hr. for a 50% probability of detection, R is 50 miles, and the speed of the intruder submarine is assumed to be 25 kts. Using these data, sprint time T_S, is calculated from formula 4.6 as 1.35 hrs. This time satisfies the conditions based on target availability and specified detection probability. Using this value and calculating V_G (formula 4.8), the speed of advance of the screen is 22.4 kts. At this rate, the screen would run away from the screened force. On the other hand, increase in search time would reduce the SOA of the screen and concommittently improve detection probability.

Increasing search time to the 87.5% probability level (3 looks) results in a screen speed of advance of 19.8 kts, a solution of practicable value under the circumstances specified.

Because of the number of variables involved in sprint and drift screening, and their interdependency, a series of tables such as presented above for the hydrofoil case would be of value in any specific screening situation. These tables could be made prior to departure for the operation, based on sprint speeds of the platforms available, the speed of advance desired, the predicted detection ranges based on sonar FOM, and an assumed target speed of penetration. If the tables included a nominal range of values each for range, search time, usable search speeds, sprint speeds, speed of advance, and assumed target speed, a baseline for accommodation to most any change in the situation would be readily available.

The hydrodynamic aspects of the towed array at sprint speeds are deserving of comment. The drag force of the towed system in the water is a function of a number of variables as indicated in the formula

$$D \text{ (drag)} = 1/2\, C_D\, \rho\, d\, V^2\, \ell \cos \phi \qquad (4\text{-}11)$$

where C_D is drag coefficient of the cable cross-section,

ρ is the density of sea water,

d is the cable diameter

V is the tow velocity

ℓ is the length of the tow cable, and

ϕ is the angle of the tow cable from the vertical.

Allowable drag for any given installation can be maintained through a range of values of cable length and tow speed. In order not to exceed the allowable drag, during sprint it may be necessary to haul in the array to short stay, or to even haul it on board. From a tactical point of view, the time to haul in and redeploy the array would be included in array stabilization time previously discussed. Towing and retrieval systems for towed arrays are normally specified in payout and retrieval capability in feet of cable per minute.

Notes

1. R. J. Urich, *Principles of Underwater Sound for Engineers* (New York: McGraw-Hill Book Company, 1967), p. 45.

5 Acoustic Considerations in ASW Operations—A Summary

Summary of Factors Affecting Sonar Tactical Performance

Table 5-1 is a summary of the factors discussed in the preceding chapters, along with an indication of who or what *controls* each factor. If one groups the controlling items from the right side of table 5-1, tactical sonar performance might be considered represented by the following summation.

> Tactical Sonar Performance = Sonar design + ship design + ship operation + sonar operation + target design + target operation + sonar condition + sonar operator training + tactical geometry + environment + target availability.

These controlling items may be further grouped as follows:

1. Items not under the control of the ASW commander:
   own sonar design
   own ship design
   tactical geometry
   environment
   target design
   target operations.
2. Items under the control of the ASW commander (ship or force):
   own sonar condition
   own sonar operator training
   own ship operation
   own sonar operation.
3. Items the ASW commander can measure:
   environment
   sonar figure of merit

Of the items listed as not under the control of the ASW commander, *own sonar design* has already been discussed and

Table 5-1
Summary of Factors Affecting Sonar Tactical Performance

Factor	Controlled by
Factors Affecting Passive Sonar Tactical Performance	
Target Signature	*Target* Design Characteristics *Target* Speed *Target* Machinery Quieting
Background Noise	
Self-Noise	*Own Ship* Speed (Flow Noise) *Own Ship* Machinery Noise Field
Ambient Noise	*Environment* (Sea State, Wind, Shipping Noise, Sea Life, Adjacent Ships in Own Force).
Directivity	*Own Sonar* Array Design *Own Sonar* Condition
Recognition Differential	*Own Sonar* Design *Own Operator* Training *Own Sonar* Condition
Propagation Loss	*Environment* (Sound Paths Available, Bottom Loss, Layer Depth, Ocean Area, Season, Time of Day)
Attenuation	*Own Sonar* Design (Frequency)
Probability of Detection	*Target Availability*
Additional Factors Affecting Active Sonar Tactical Performance	
Source Level	*Own Sonar* Design, Power Available
Target Strength	*Target* Aspect *Target* Reflectivity
Reverberation	*Environment* (Sea Structure) *Own Sonar* Source Level

analyzed in preceding chapters. The sonars, as they are designed, fabricated, and delivered are basically a non-variable to the ASW commander when considered in terms of design alone. Likewise, *own ship design* is basically a fixed factor to the ASW tactical commander. He cannot change its basic speed characteristics, nor the machinery noise radiated as a function of its speed. Neither can he change the flow noise over the sonar hydrophones, which are a fixed installation in his vessel(s). Likewise, the *target ship design,* its radiated noise characteristics, the quieting measures that may have been utilized in its construction, whether or not anti-reflective coatings have been installed to reduce target echo strength, *the skill with which the enemy submarine is operated, the alertness and training of the enemy sonar operators;* all are factors beyond the direct control of the tactical ASW commander. But the ASW commander needs to have an awareness and appreciation of the capability of the prospective enemy submarine to operate his ship and sonar in a manner best suited to prevail in the tactical encounter that may ensue. The *tactical geometry* of the encounter may result from chance or from a priori intelligence on either side. If an equal sonar design capability is accorded to each side in the encounter, then he who best understands and exploits the ocean environment and the acoustic operational factors attendant thereto will be expected to prevail.

The items under the direct control of the ASW commander include the material *condition of his own sonar(s),* the *state of training of his sonar operator(s),* the manner of *operation of his own ship(s),* and the *operation of his own sonar(s)* as a matter of policy.

Material condition of the sonar and state of operator training make an obvious and controlling contribution to own sonar capability; a control so obvious that no further discussion is meaningful within the context of this text. The sonar and the operators either have been made ready for their mission, or they have not.

With respect to sonar operation once at sea, the commander has the capability to *measure the ocean environment,* and to interprete that environment in the light of his own mission and the predicted performance of the sonar assets available to him. In this connection, a review of the variability that may be expected is useful.

Figure 5-1 is a summary of the parameters of active and passive sonars as contained in the sonar equation. In normal use, these parameters are combined to compute a figure of merit (FOM) for the sonar system, which in turn is used to enter a propagation loss curve for a detection range prediction. For most of the parameters in the sonar equation, while a specific nominal value is used in computing FOM, it should be recognized that variability in each parameter creates a range of variability or uncertainty. N_{DI} is the one parameter that is relatively invariant for both active and passive sonar systems.

It has already been stated that the variability of N_W is on the order of 5-6 dB. It has also been stated that values of N_{TS} normally range from 10 dB for a bow or stern target to 25 dB for a beam target. The nominal use of 15 dB creates another variability on the order of ±7

PARAMETER	ACTIVE SONAR		PASSIVE SONAR
	NOISE LIMITED	REVERB LIMITED	
SONAR SYSTEM	L_S N_{DI} N_{RD} L_N (SHIP)	L_S N_{RD}	N_{DI} N_{RD}
OCEAN MEDIUM	$2\ N_W$ L_N (AMBIENT)	$2\ N_W$ L_R (REVERB)	N_W L_N (AMBIENT)
TARGET	N_{TS}	N_{TS}	L_S

Figure 5-1. Basic Parameters of Active and Passive Sonars

dB. Sea state and thus L_N (ambient) can vary through a range of values of about 25 dB. Performance predictions made during planning for one sea state should be adjusted appropriately to the sea state actually observed during the operation. And for active sonars, reverberation conditions may change daily as a function of time of day.

There can also be a range variability as a result of the sonar operator changing his false alarm rate. Assume the operator at the beginning of the operation was a little too quick in his detection decisions, resulting in a high false alarm rate (and a low value of N_{RD} at the 50% probability detection range). Subsequently, to improve his false alarm rate, he became more methodical and used more time for each detection decision. This adjustment, while reducing his false alarm rate, also increased the numerical value of N_{RD}, which in effect reduces FOM and thus the 50% probability range prediction.

For active sonars, L_S is normally relatively invariant for a given transmitter power setting. On the other hand, L_S for passive sonars may be expected to be target-speed dependent to some extent. Depending on information available as to potential target radiated noise characteristics vs. speed, a realistic value of L_S for the sonar equation could be obtained by assuming a target speed adequate to accomplish the submarine's mission.

Some appreciation can be made for the relationship between the standard deviation assumed in both FOM and N_W as related to

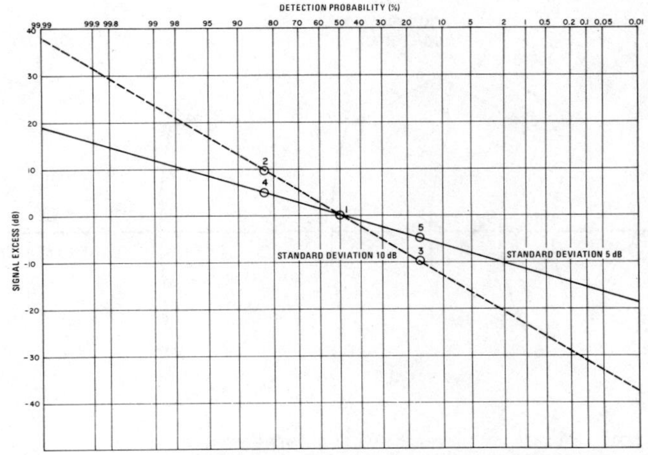

Figure 5-2. Detection Probability for a Given Range as a Function of Signal Excess for Standard Deviations of 10 dB and 5 dB Respectively

detection range and probability of detection by making detection range contour plots. This involves calculating FOM, constructing a signal excess vs. probability of detection diagram, and then using a propagation loss curve in the following manner.

Assume an active sonar with FOM of 196 dB, operating in the Norwegian Sea at the frequency specified for figure 3-10. Further, assume that the composite standard deviation from these nominal values for all the parameters discussed above, including N_W, is ± 10 dB. On probability paper, establish a signal excess scale as in figure 5-2 and plot points 1, 2, and 3 in the following manner. Point No. 1 represents a probability of detection of 50%. By definition in the sonar equation, this occurs where the signal level just equals noise level—when the signal to noise ratio is zero. (When the signal to noise ratio is zero, there is no signal excess.) Point No. 2 is located at 10 dB (1 standard deviation) from the zero signal excess axis at the 84% probability line. Point No. 3 is similarly located, only at the 16% probability line. A line connecting these three points relates detection probability to signal-to-noise ratio.

The allowable one-way propagation loss for any probability of detection of interest is calculated by subtracting from the sonar FOM for passive sonars (or 1/2 FOM for active sonars) the appropriate signal excess from figure 5-1, for the standard deviation as-

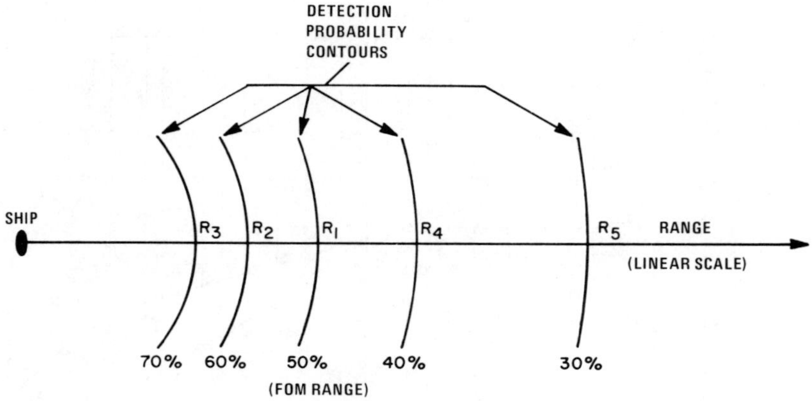

NOTE: PROBABILITY CONTOURS WILL BE UNEVENLY SPACED BECAUSE PROPAGATION LOSS IS NOT LINEAR WITH RANGE.

Figure 5-3. Detection Probability Contours as a Function of Range

sumed. For example, for the sonar projected above, the allowable propagation loss for a 70% probability of detection assuming a 10 dB standard deviation is 196/2 - 6 or 92 dB. The corresponding range from figure 3-10 is as follows:

Direct Path	*Bottom Bounce*	*Conv. Zone*
7.8 kyds.	—	60 kyds.

In a similar manner, the predicted range for any probability of detection can be computed. By so doing for a range of probabilities such as 30%, 40%, 50%, 60%, and 70%, and then plotting these ranges to scale from a point of origin representing own ship, a plan view of detection range vs. probability of detection is available, as illustrated by Figure 5-3.

Detectability of the ASW Force

A factor significant to the ASW encounter is the acoustic portrait of own ship, as portrayed to the enemy submarine. This acoustic portrait has two major dimensions; that of own ship radiated noise signature, and that of own ship active sonar emissions. These are discussed below.

The capability of the opposing submarine to detect the screen and the screened force is a consideration, which, when coupled with assumptions of the capability of the opposing submarine weapon system, will be significant in the selection of the tactical ASW plan. While a complete analysis of opposing weapons systems is beyond the scope of this book, an understanding of the acoustic factors affecting the capability of the opposing sonar to detect own forces is in order. These factors are summarized in the list below.

Factors Affecting Own Ship Acoustic Detectability by the Opposing Submarine

Factor	*Controlled by*
Own Ship Radiated Noise	*Own Ship* Speed, Quieting
Acoustic Emission Control	*Own Ship* Policy
Target Sonar Performance	*Target* Sonar Capability
Deception	*Own* Acoustic Devices to Mask Signature
	Own Screw Masking

Own Ship Radiated Noise

This includes noise from main propulsion and auxilliary machinery, and from hydrodynamic sources. The contributing sources of radiated noise are as indicated in figure 5-4. These different noise sources produce different types of frequency spectra, and the resulting spectra of an individual ship are a combination of a broad continuous spectrum through a wide frequency range, and single frequency components called "line" components. Examples of various ship spectra have already been shown in figure 3-23. It is not the intention here to discuss in detail or even define all of the sources of ship's radiated noise listed in figure 5-4. It is intended that the reader understand that any and all shipboard machinery contributes to the signature portrait provided to the enemy submarine.

While the radiated noise signature of a ship is largely fixed by its design, radiated noise does vary between ships of the same design and according to speed in individual ships. A typical destroyer several years ago demonstrated the following variations with speed.

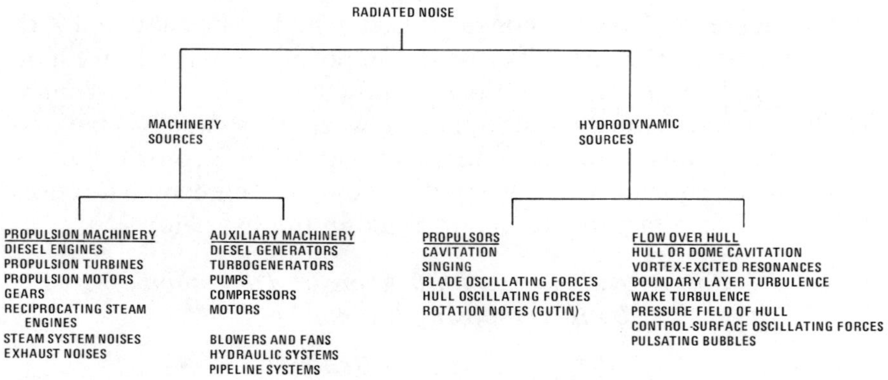

Figure 5-4. Sources of Noise Radiated by Surface Ships, Submarines, and Torpedoes

	Frequency Band 2-17 Hz			
Speed (kts)	5	10	15	20
Spectrum Level, dB	52	55	63	73

These numbers demonstrate that one measure of control over ship detectability is in selection of speed.

A second means of obtaining control over own ship radiated noise is to periodically measure the ship's noise signature on an acoustic range to determine changes that may have occurred and to record the effect of different combinations of machinery at different speeds. Least noisy combinations of equipment can be selected when feasible.

A third way of controlling ship radiated noise is to modify the manner in which machinery sound is transmitted to the hull. Such modifications, called *ship quieting* measures, are usually only susceptible of implementation during shipyard or tender overhauls.

A fourth way of controlling radiated noise is by the use of hull damping coatings.

For the tactical commander, the means normally available to reduce own ship radiated noise levels is through control of speed and combinations of machinery in use. This requires some knowledge of the radiated noise signature of own ship, acquired through a systematic program of measurement.

Sonar Acoustic Emissions

Understanding of the sonar equation for active sonars should point out the fact that active transmissions are detectable at extremely long ranges compared to the range at which a detection may be predicted. Consider the situation where own sonar has a figure of merit of 196 dB, including a sonar source level of 140 dB. Since two-way propagation loss is involved, the one-way allowable loss is 98 dB. Using the propagation loss from figure 3-10, the active ranges predicted are:

Direct path	9.4 kyds.
Bottom Bounce	48.0 kyds.
1st CZ	58.0 kyds.

Now consider the energy put into the water, and the energy level at the target.

LS	Sonar radiated signal level	140 dB
NW	one-way propagation loss	98 dB
	Signal Level at the target	42 dB

If it is assumed that 10 dB of signal energy is absorbed by spreading and attenuation each time the range is doubled, for the bottom bounce and convergence zone paths, the signal level would reduce to zero at about the following ranges:

Bottom Bounce	16 × 48 kyds.	= 768 kyds. (384 miles)
Convergence Zone	16 × 58 kyds.	= 928 kyds. (464 miles)

Therefore, one could say that if the opposing sonar had the equivalent of own sonar processing gain, it would detect our presence at a distance significantly greater than 16 times our own detection capability against it.

To be realistic, it should not be assumed that the opposing processing gain is the equivalent of our own, since this also assumes the enemy has detailed knowledge of our own signal format, which is operator variable in most modern active sonars, and which to the interceptor is a transient phenomena. Nevertheless, the point can be made that a counter-detection advantage against an active sonar exists and that its magnitude easily ranges from 2 to 8 times, or more, our own capability prediction against the submarine.

Acoustic Emission Control Policy

The potential detection advantages afforded the enemy submarine through the radiated noise signature of own ship(s), and the detectability of own active sonar emissions warrants the serious consideration of a policy to limit these emissions, and/or to use them to create an ambiguous acoustic picture to reduce or equalize whatever long-range detection advantage the submarine may have by requiring him to close own force to resolve his uncertainties. Some of the ways to accomplish this include screw count masking, (varying the revolution count of the screw(s) to make more difficult enemy "count" of screw revolutions, and thus estimating speed) and the use of acoustic devices to obscure or mask own acoustic signature. Other ways might include the use of sprint and drift tactics to vary the composite radiated noise signal levels own force generates. Still another might include use of a formation spread in both front and depth of the force at sea. Figure 5-5 illustrates the irregular width of the acoustic target presented as a function of distance of the submarine from the transiting force, and the width of the formation in miles. If a part of this width were obtained through use of acoustically "synthetic" ships, the effectiveness of long-range weapon systems might be significantly blunted.

The Tactical Environment

As has been previously stated, the environment is variable, which contributes to the degree of difficulty of the ASW detection problem. As counter to its variability, the velocity profile of the environment is measurable. Granted it is measureable to both the hunter and the hunted, the understanding and exploitation of that which is measured will probably determine the victor, or at the minimum,

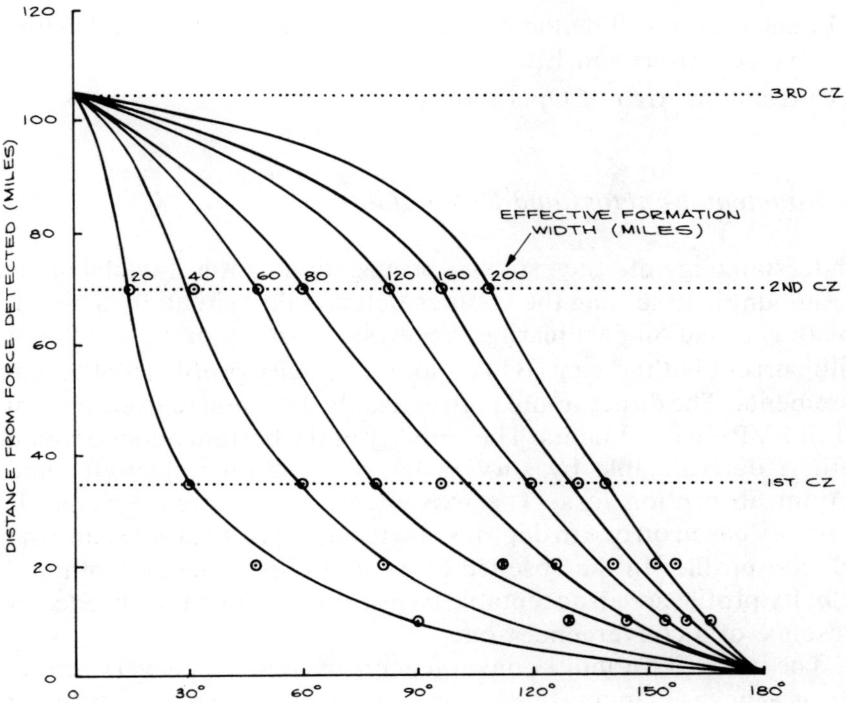

Figure 5-5. Angular Width of Force as a Function of Range and Force Width, Assuming Original Detection, Made at 3rd Convergence Zone

determine who holds the edge in detectability vs. counter-detectability in the encounter. It is generally accepted that the advantage belongs to he who gains the initial detection in the ASW encounter.

The list below contains some of the measurable factors which will affect own ship detectability by the enemy, and own ship capability to detect him.

Environmental Factors Affecting Detectability and Counter-Detectability

Sound Paths Available
Shallow Water vs. Deep Water
Layer Depth
Seasonal Variation in Area (Wind, Temp, etc.)

Local Transient Phenomena (Rain, Passage of Cyclonic Disturbance, Afternoon Effect)
Currents in Area of Operations.

Environmental Factors and Detectability

Understanding of the sound propagation paths available is paramount in assessing the counter-detectability situation. This estimate is based on past marine geophysical survey data, correlated with current bathymetry (BT) or sound velocity profile (SVP) measurements. The direct sound path is readily determined from current BT or SVP measurements. The viability of the bottom bounce sound path is determinable by survey data on bottom reflectivity and bottom absorption loss. The existence of convergence zones is normally based on ocean depth, knowledge of prior surveys, and the velocity profile. In the absence of prior surveys, ocean depth and velocity profile are an acceptable basis for predicting the absence or presence of a convergence zone.

The concept for multi-convergence zone contacts is very attractive as a means of increasing sonar search rates, and may be brought to fruition by the long towed linear array. However, convergence zone detections raise serious problems of target availability as a function of the width (or depth) of the convergence zone annulus and the target speed of penetration of the annulus. This in turn creates tactical problems for the operational commander in target localization, and in subsequently prosecuting the attack, or in evading the submarine. Tactics need development to assist in answering such questions as how to track effectively a target penetrating through multi-convergence zone annuli, or how to determine in which convergence zone the target is located.

With respect to detection in shallow water, it has already been indicated that sonar performance may be enhanced because the ocean bottom and surface boundaries act as a duct for the channeling of sound. Since no precise quantitative measure of the expected improvement is available, the ASW tactical commander could view the shallow water problem as one in which ambiguities may be created by him to mask the real composition and precise presence of his forces. One can even conceive of his active sonars being used in a bistatic mode, i.e., using one ship's sonar as an ensonifying source, and any other ship's sonar or sonobuoys as receiving or detecting sensors. Bistatic geometry creates problems in establishing precise target location. On the other hand, indication of target presence may tip the tactical balance, or at least provide the tactical commander with alternative courses of action, no matter how target presence was established.

The depth of the bottom of the surface layer of water is a great determinant of sonar performance from a hull mounted surface ship sonar because the submarine may choose to proceed below the layer. As was discussed earlier, cross-layer detection is usually limited in range because of the refraction or bending of the sound rays. The tactical answer to this situation, from the escort point of view, is to vary the vertical angle of transmission of his sonar projector so as to penetrate the layer, or to deploy a variable depth sensor below the layer.

The predicted seasonal variation in the area of operations based on previous surveys and observations of wind action and temperature profile is basic data for the staff operational planner. When the force is "in situ," the tactical commander should validate or verify the predictions by regular periodic measurements of the ocean temperature or sound velocity structure. The length of the period between measurements should be based on observations of current weather phenomena affecting sound propagation conditions, i.e., wind, time of day, etc. It is particularly important to be aware of the surface heating effect of the sun and the fact that in areas of little mixing of the sea by wind driven waves, a positive temperature gradient may be developed in the surface layer between mid-morning and mid-afternoon which seriously degrades surface ship sonar performance. Under these conditions, hourly measurements of the environment are in order. Similarly, when conducting operations in the vicinity of currents like the Gulf Stream or the Laborador Current, where "fingers" of water of marked difference in adjacent waters is common, the tactical commander should consider hourly measurements.

The Sonar and Weapon Delivery

As a sensor in the ASW weapon system, the task of the sonar is to provide target position information with sufficient accuracy for effective weapon placement. Considering first the active sonar, one can measure directly, range and bearing. Modern active sonars are capable of accurate range and bearing measurements. However, when compared with radar, the rate of updating previously measured target positions is very, very slow. This slow data rate, resulting from the comparatively slow speed of sound in water, adversely affects the capability of the ASW fire control system to establish and maintain a smooth target track as the basis for predicting target position and thus weapon placement. This adverse effect of the properties of sound in water must be overcome by doctrinal practices beyond the scope of this paper.

The passive sonar presents a different problem in target localization with sufficient accuracy for weapon placement, because the passive sonar only measures directly target bearing. Calculation of target range, as well as course and speed for predicting the future position of the target, and subsequent weapon placement, is a matter of tactical doctrine involving various ship maneuvers, beyond the scope of this book. Because of the need for multiple maneuvers, the time required to obtain a fire control solution suitable for weapon placement is greater than that normally expected with an active sonar.

Appendix A
Some Conversion Factors and Scales for Converting Between Metric and English Units of Measure.

Useful conversion factors and scales.

1 foot = 0.341 meters; 1 meter = 3.281 feet

1 yard = 0.914 meters; 1 meter = 1.094 yards

1 nautical mile = 1.850 kilometers; 1 kilometer = 0.540 nautical miles

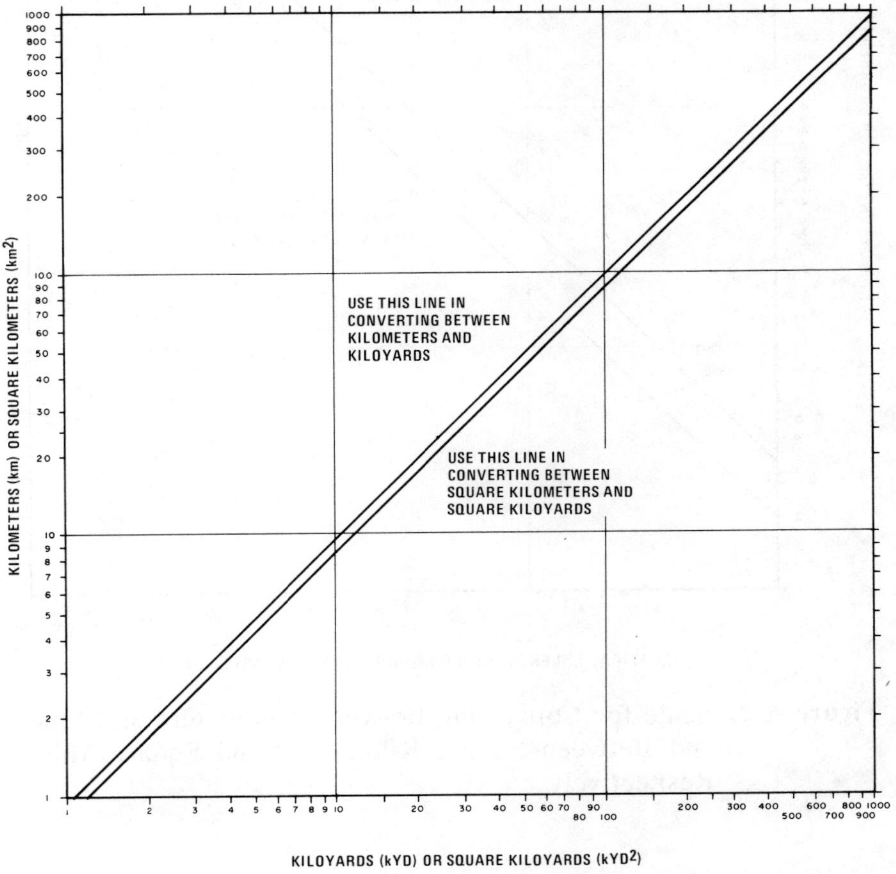

Figure A-1. Scale for Converting Between Kiloyards and Kilometers and Between Square Kiloyards and Square Kilometers, Respectively.

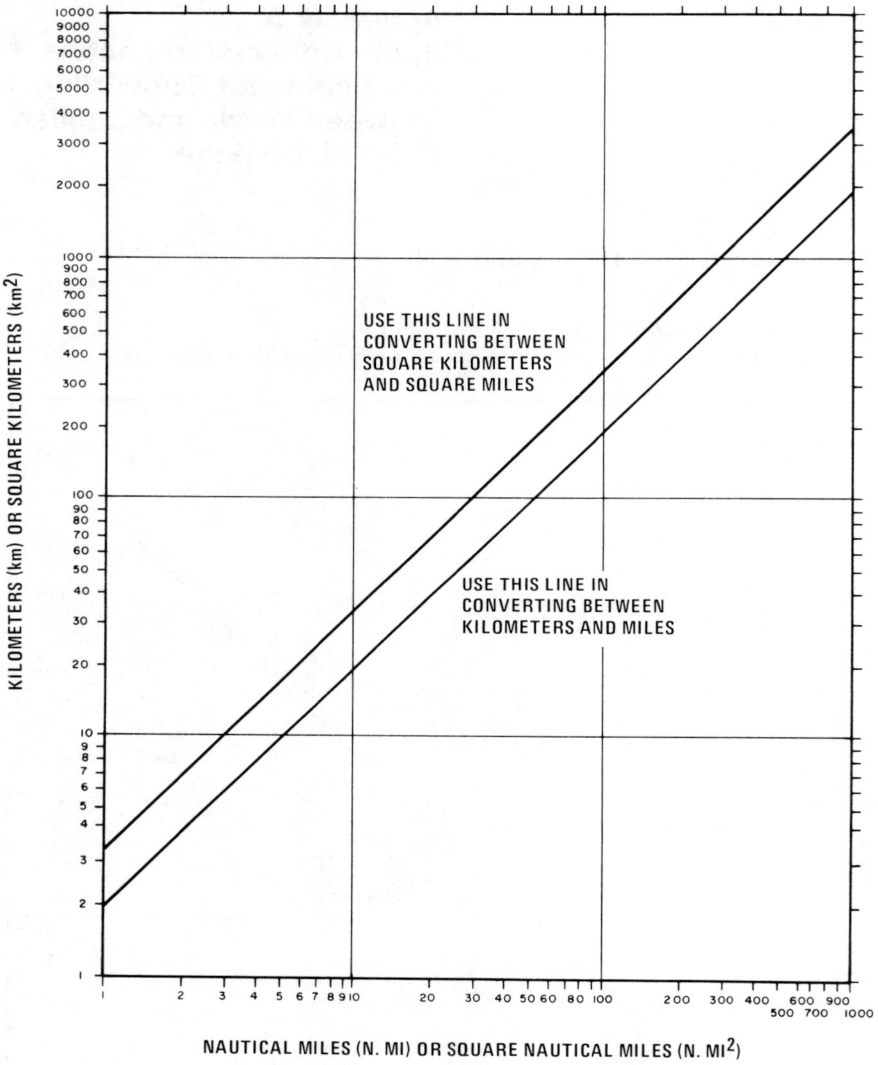

Figure A-2. Scale for Converting Between Kilometers and Miles and Between Square Kilometers and Square Miles, Respectively

Appendix B
Compendium of Useful Working Graphs

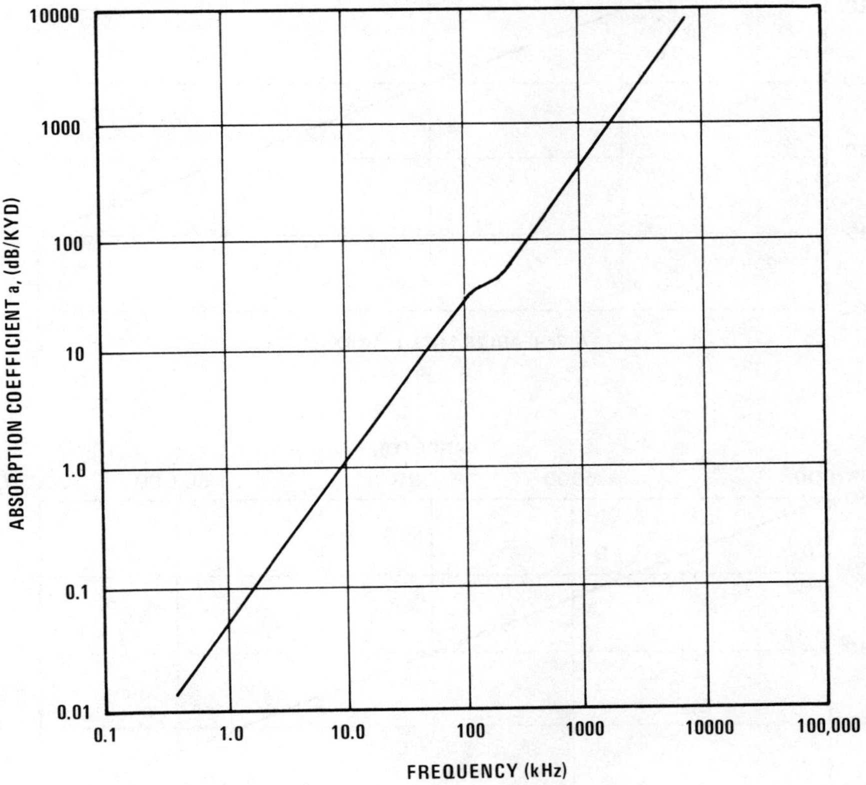

Figure 2-6. Absorption Coefficient in Sea Water of Salinity 35 ppt at 39° F. as a Function of Frequency

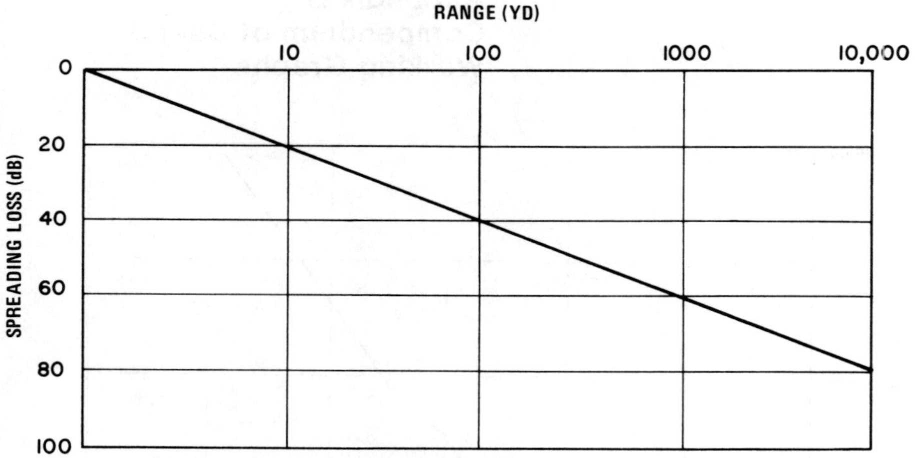

A. SPREADING LOSS, 1–10,000 YDS

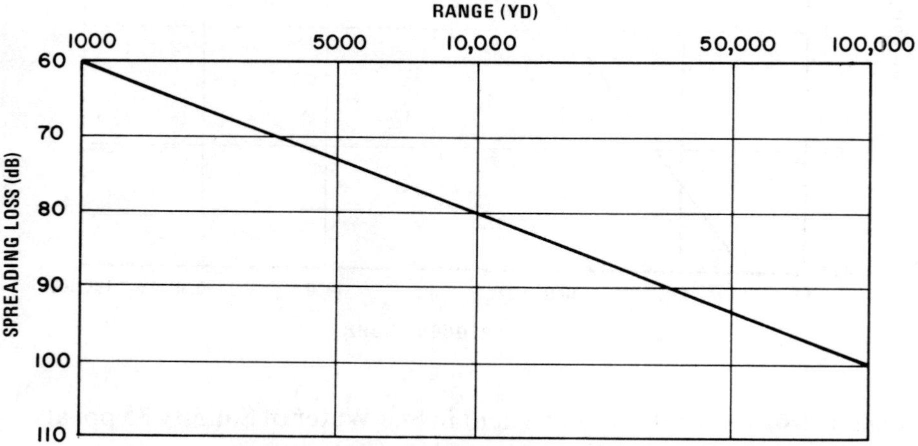

B. SPREADING LOSS, 1000–100,000 YDS

Figure 3-1. Spreading Loss as a Function of Range

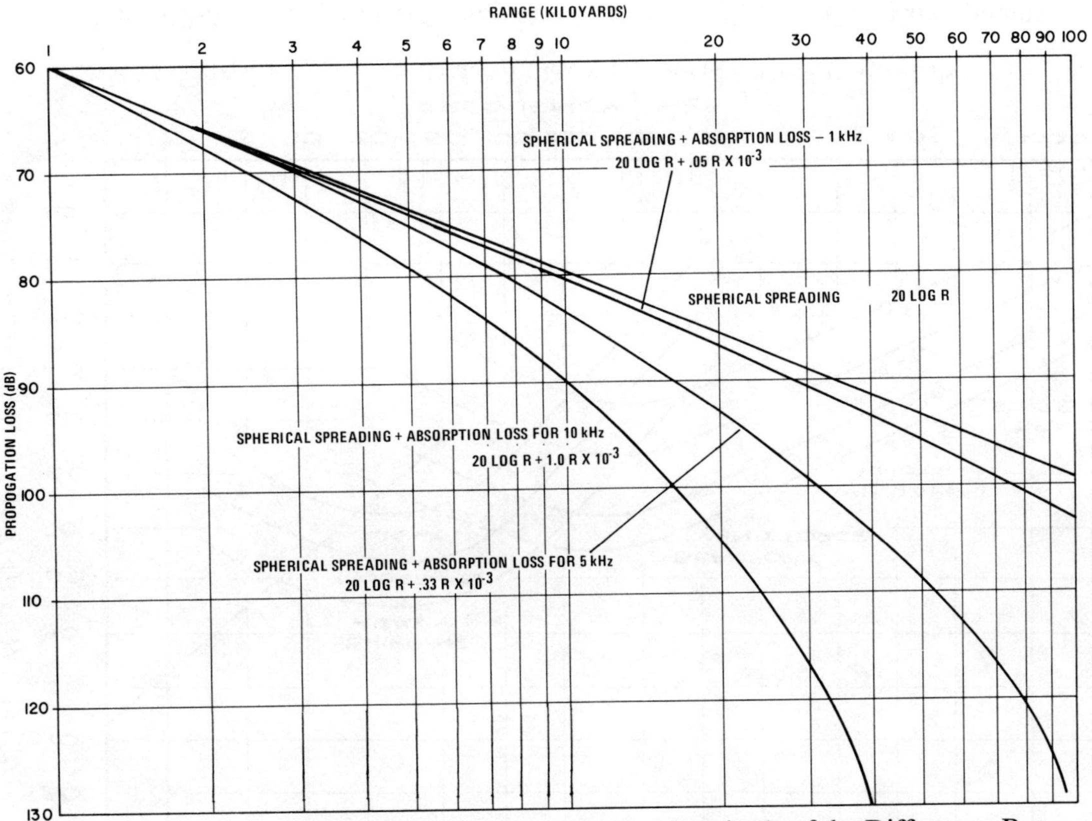

Figure 3-2. Plot of Range vs. Propagation Loss, Showing the Magnitude of the Difference Between Spherical Spreading Loss Alone, and Spherical Spreading Loss Plus Absorption Loss

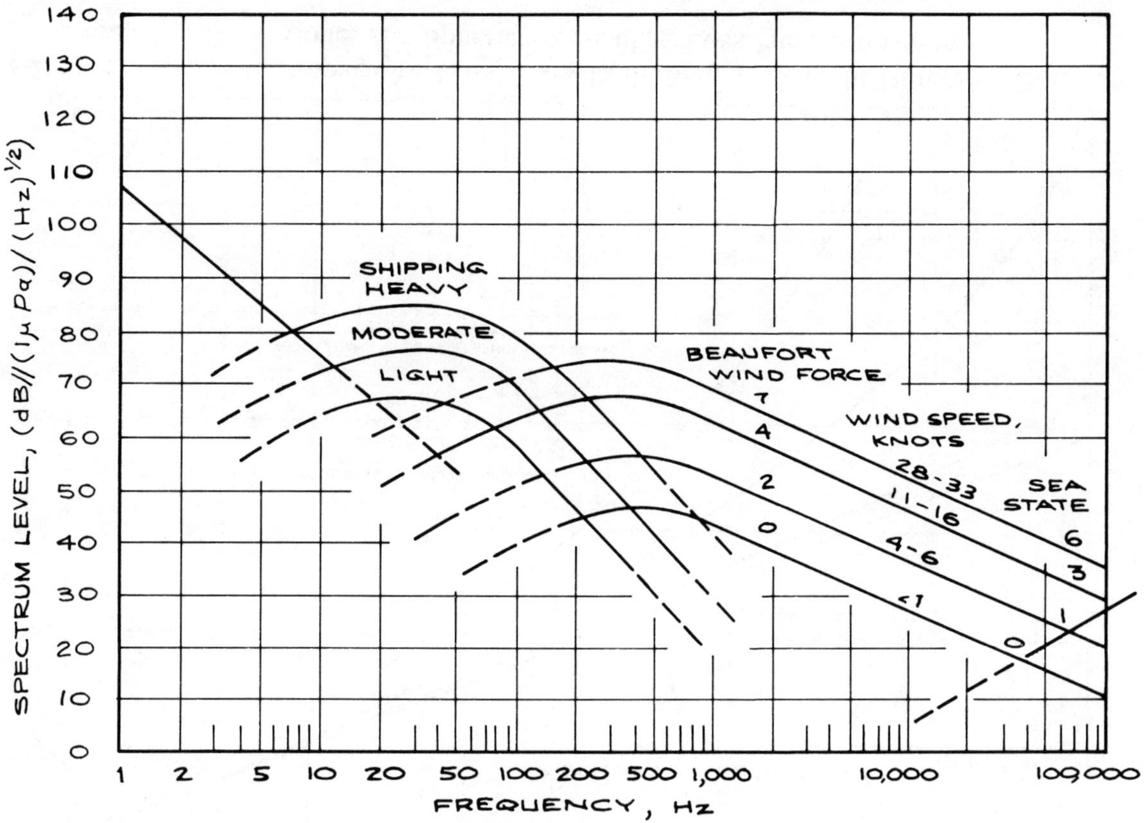

Source: R.J. Urich, *Principles of Underwater Sound for Engineers* (New York: McGraw-Hill Book Company, 1967), figure 7.5, p. 168.

Figure 3-3. Average Deep-Water Ambient-Noise Spectra (LN Component)

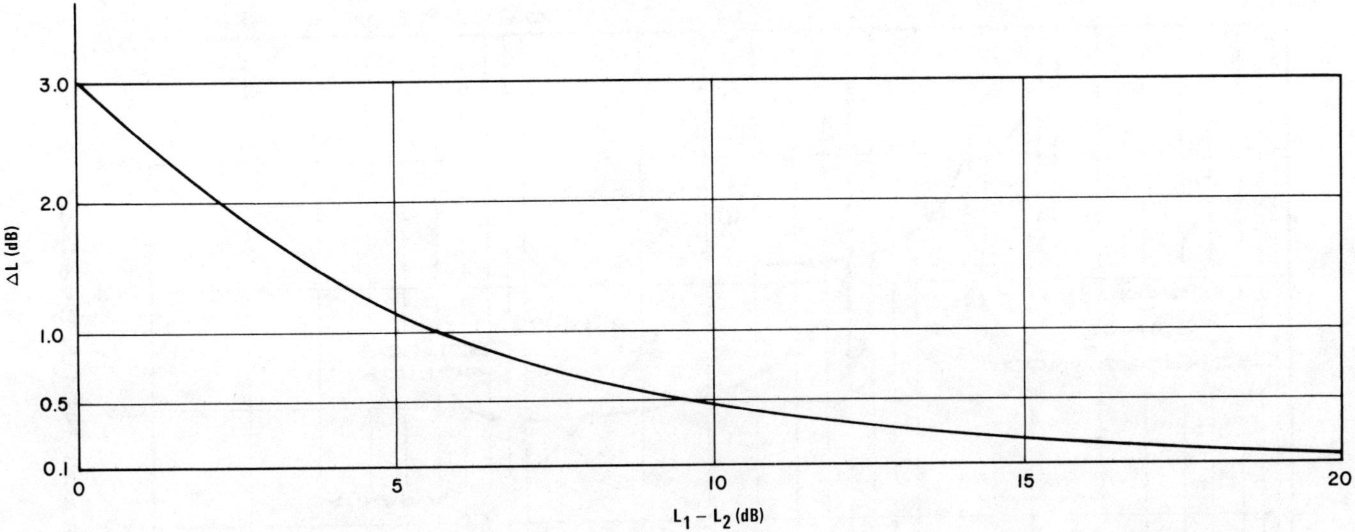

Figure 3-6. Power Summation Curve. L_T (the sum of two power quantities $L_1 + L_2$) = $L_1 + \Delta L$, where $L_1 \geqq L_2$.

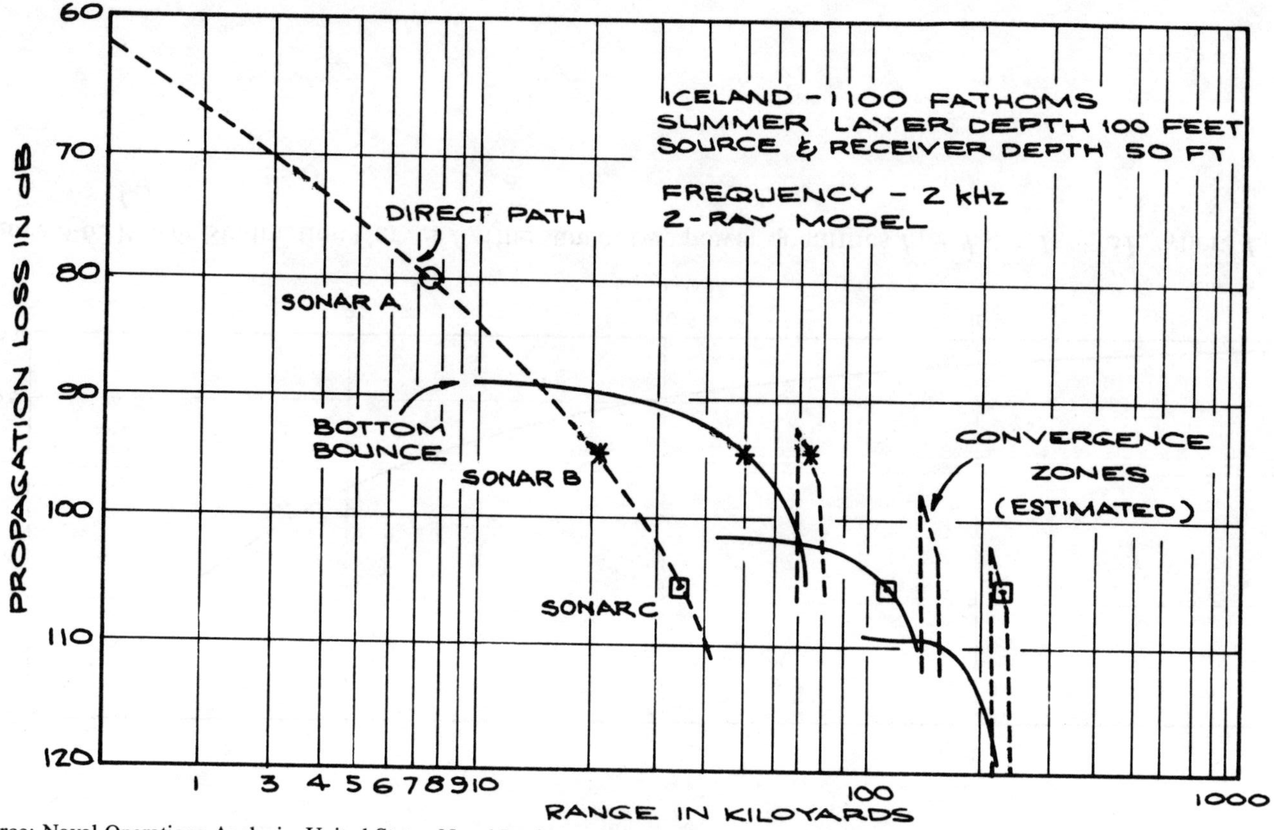

Source: Naval Operations Analysis, United States Naval Institute, 1968, fig. 9-8, p. 154. © 1968, United States Naval Institute.

Figure 3-9. Propogation Loss for the North Atlantic Ocean off Iceland in the Summer for 2 kHz

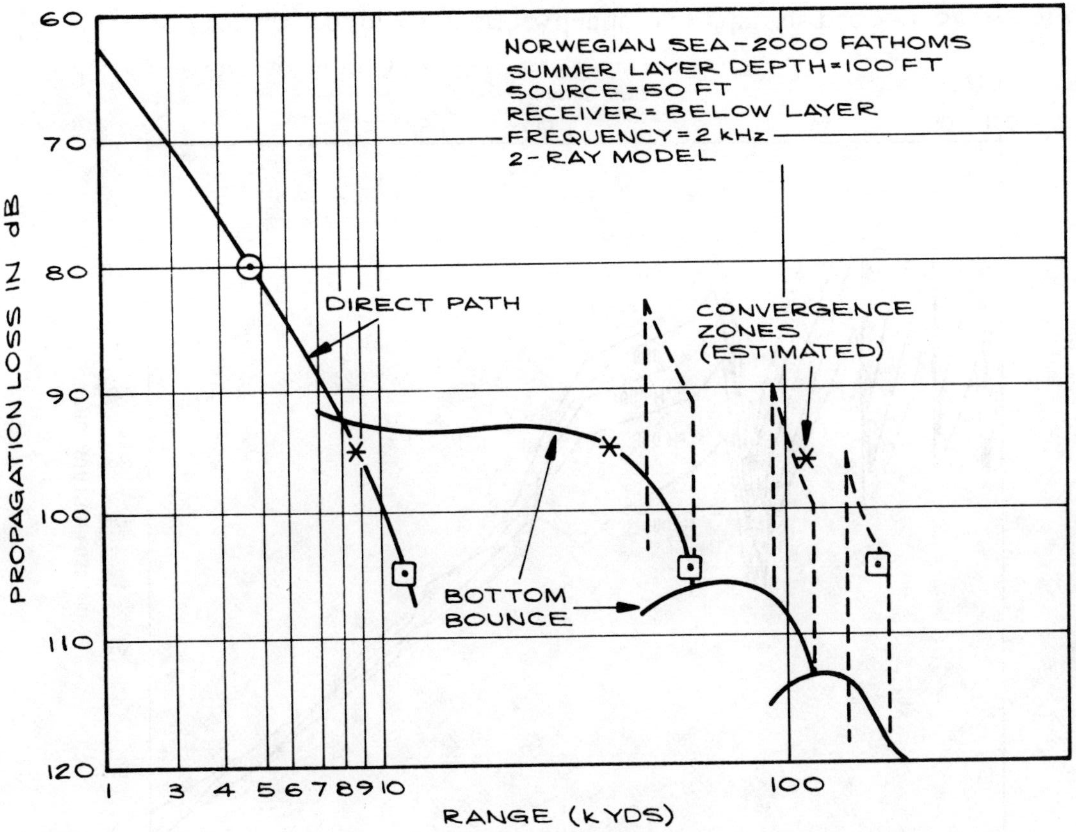

Figure 3-10. Propogation Loss for the Norwegian Sea in the Summer for 2kHz

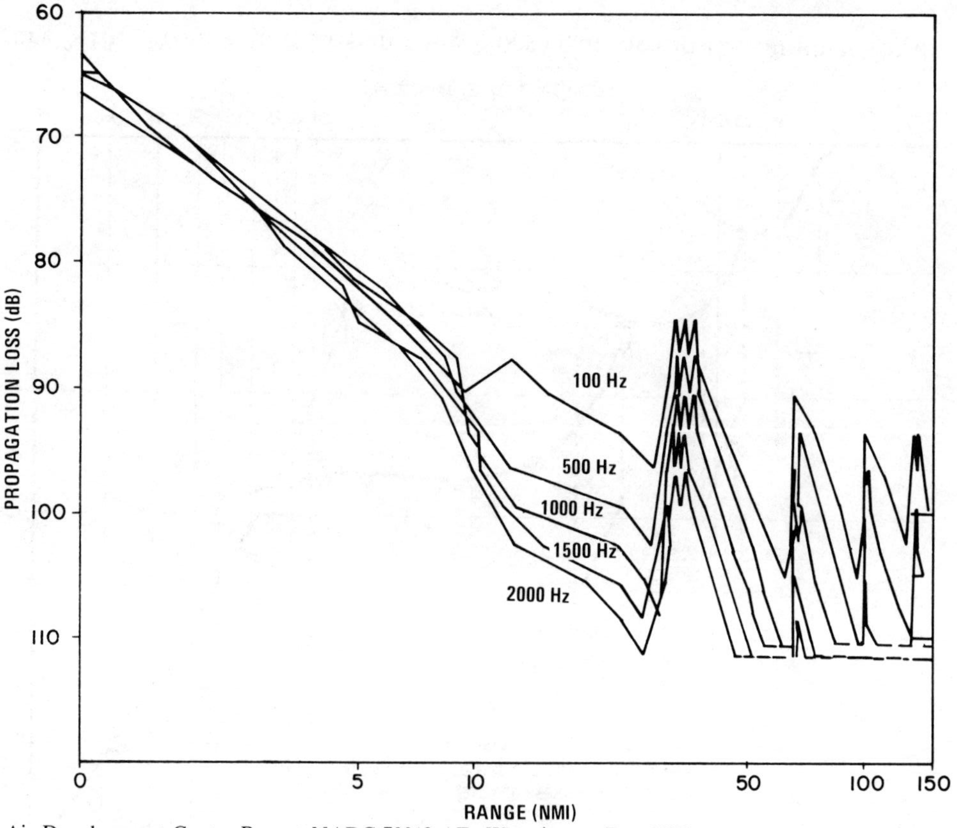

Source: U.S. Naval Air Development Center Report NADC-72012-AE, Warminster, Pa., 1972.

Figure 3-12. Average Propagation for the North Atlantic, Jan.-Mar. as per SVP Shown in Figure 3-11

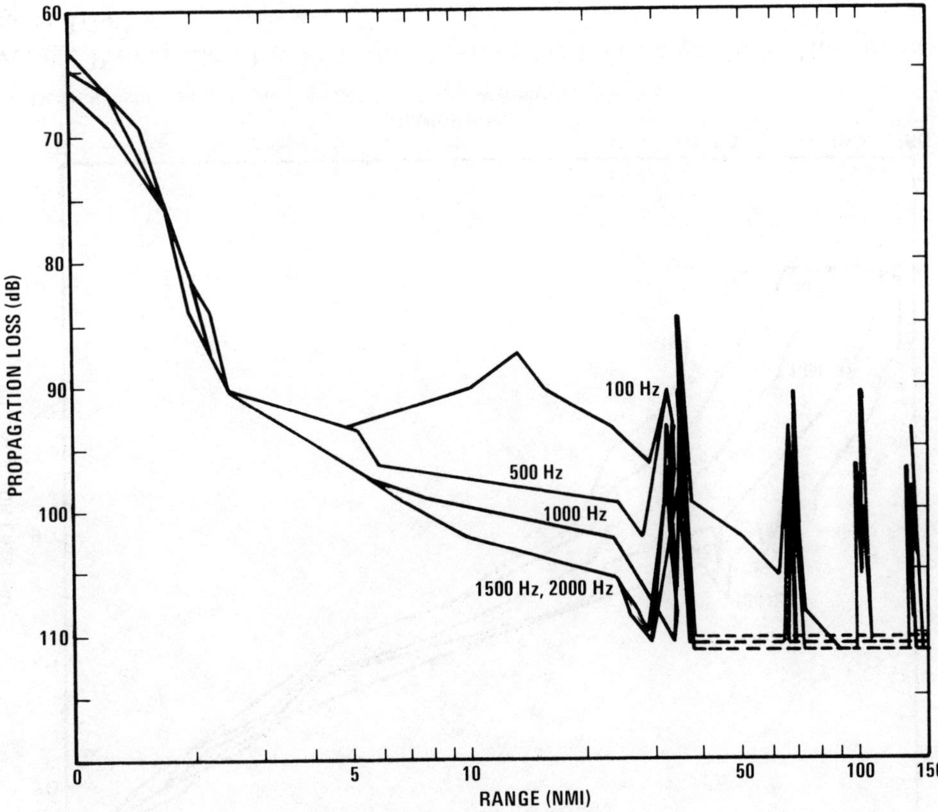

Source: U.S. Naval Air Development Center Report NADC-72012-AE, Warminster, Pa., 1972.

Figure 3-13. Average Propagation Loss for the North Atlantic, July-Sept. as per SVP Shown in Figure 3-11

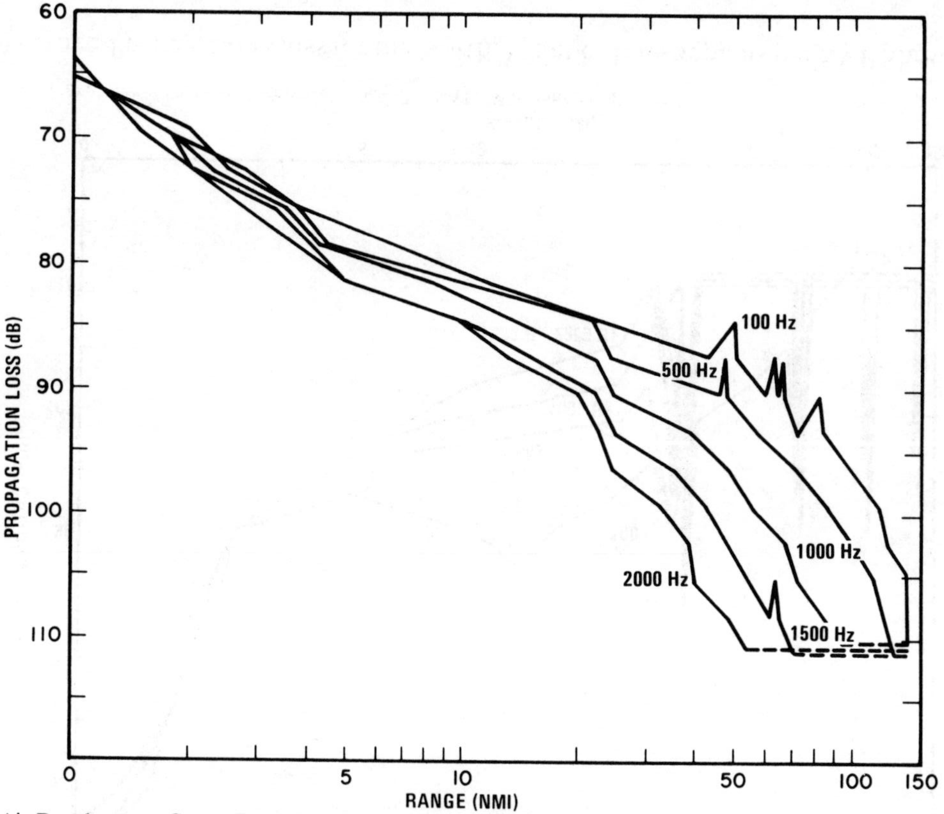

Source: U.S. Naval Air Development Center Report NADC-72012-AE, Warminster, Pa., 1972.

Figure 3-15. Average Propagation Loss for the Western Mediterranean, Jan.-Mar. as per SVP Shown in Figure 3-14

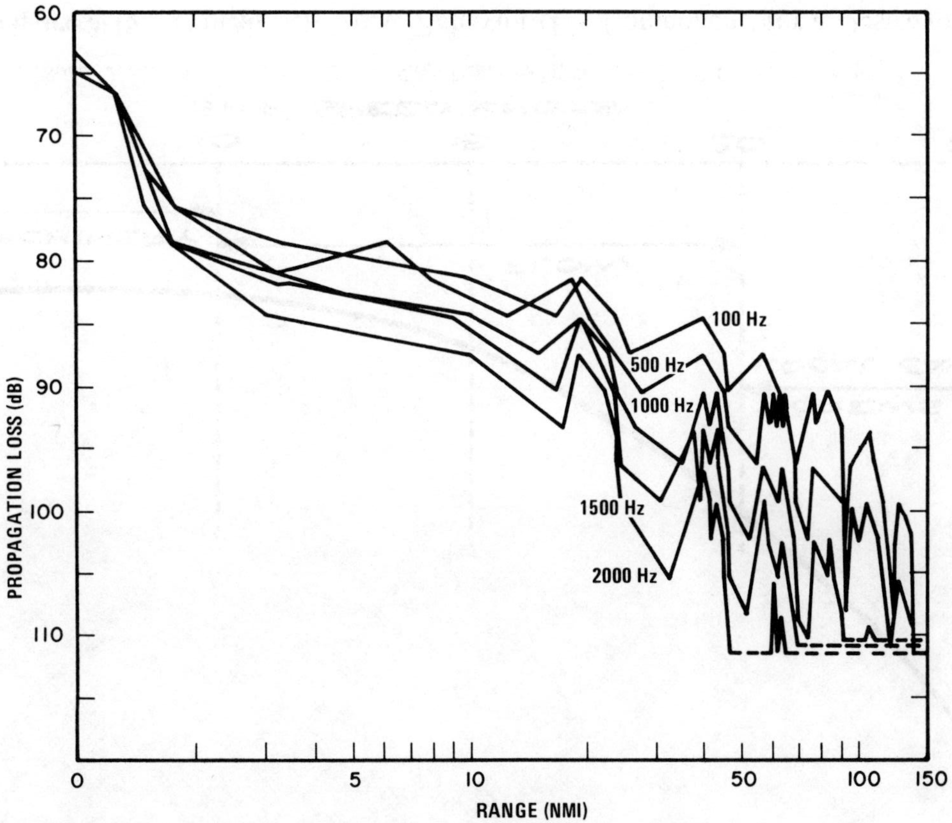

Source: U.S. Naval Air Development Center Report NADC-72012-AE, Warminster, Pa., 1972.

Figure 3-16. Average Propagation Loss for the Western Mediterranean, July-Sept. as per SVP Shown in Figure 3-14

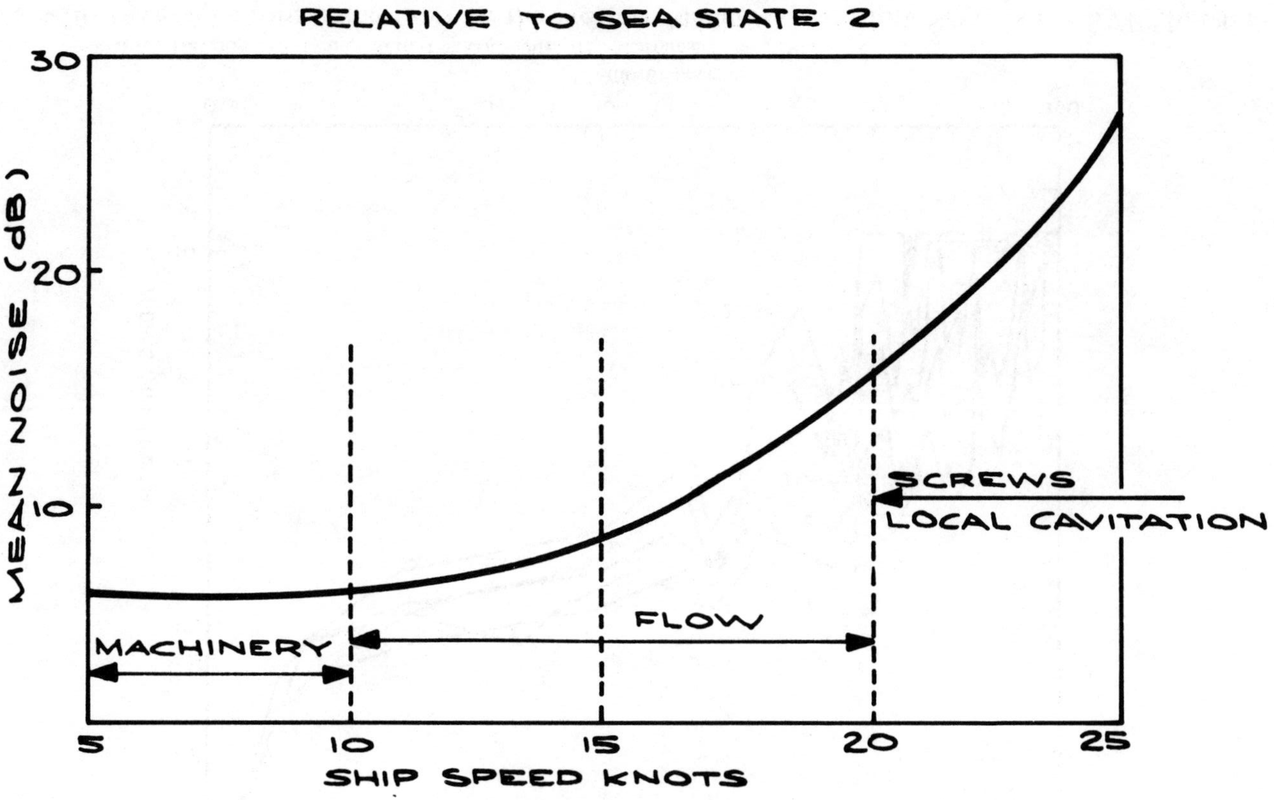

Source: *Introduction to Sonar Technology*, Bureau of Ships, Navy Department, Washington, D.C., fig. 4-7, p. 92.

Figure 3-18. Typical Self-Noise Behavior (LN Component) for a Destroyor

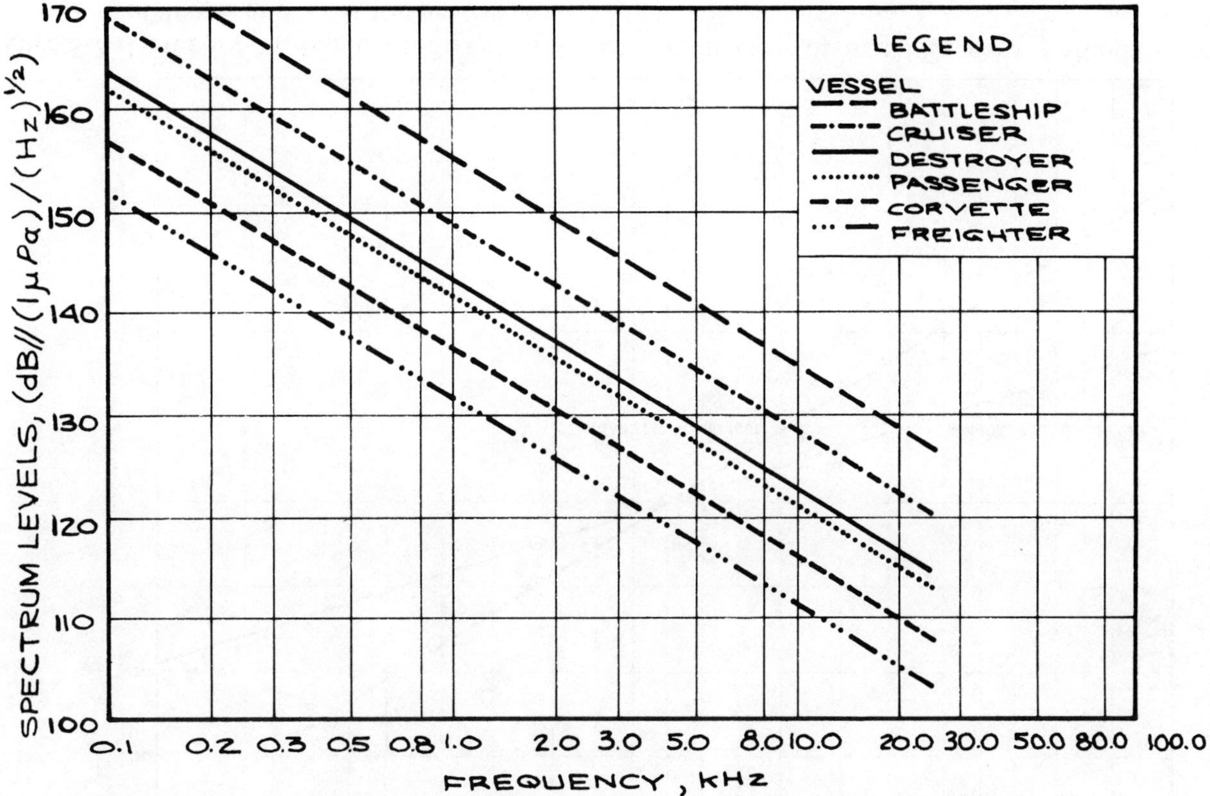

Source: *Introduction to Sonar Technology*, Bureau of Ships, Navy Department, Washington, D.C., 1965, fig. 2-3, p. 33.

Figure 3-23. Average Radiated Noise Levels of Six Different Classes of Ships (L_S)

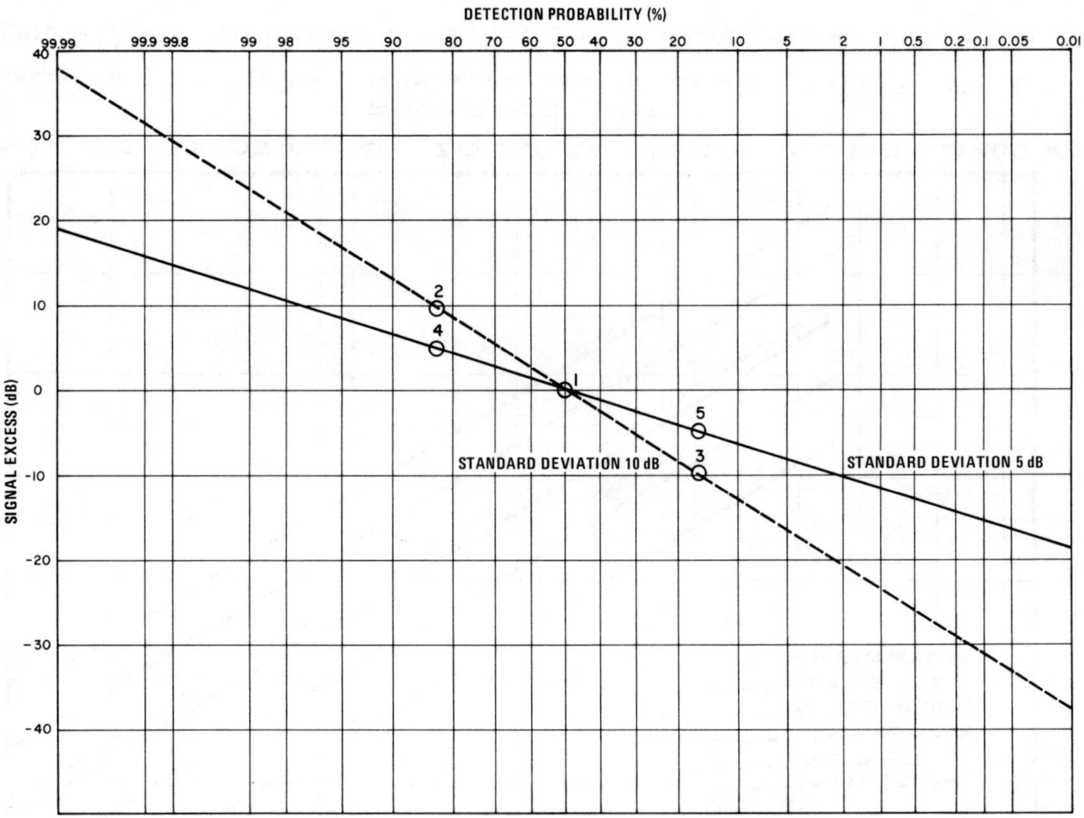

Figure 5-2. Detection Probability for a Given Range as a Function of Signal Excess for Standard Deviations of 10 dB and 5 dB Respectively

Index

About the Author

Albert W. Cox retired from active naval service in 1969 in the grade of Captain upon the completion of 27 years of commissioned service, and joined the Raytheon Company, Submarine Signal Division, where he currently is Chief Operations Analyst. During his active naval service, he commanded the following destroyer units: USS Herbert J. Thomas (DD833), USS F.M. Robinson (EDE220), USS Willis A. Lee (DL-4), and Destroyer Division 202. His extensive anti-submarine warfare experience at sea included operational evaluation of the U.S. Navy's first Variable Depth Sonar, Technical and Operational Evaluation of the U.S. Navy's first AN/SQS-26 sonar, operational evaluation of the MK 43 and MK 32 torpedoes, operational evaluation of USS Nautilus (SSN571), and Command of several U.S. and NATO Anti-submarine Screening Forces. His final tour of duty in the Navy was in the capacity of Technical Director, Anti-Submarine Warfare Systems Project Office, Office of the Chief of Naval Material, Navy Department, Washington, D.C.

Captain Cox's educational background includes the B.S. degree from the U.S. Naval Academy (1942), and graduate work in engineering management at George Washington University and in underwater acoustics at Catholic University. He is a graduate of both the Command and Staff Course and the Naval Warfare Course at the U.S. Naval War College, Newport, R.I. In addition, he completed the following Navy service schools: Fleet Sonar School, Key West, Florida; Systems Analysis Course, Navy Management Logistics School, Washington, D.C.; and the Command Course, Joint Tactical Anti-Submarine School, Londonderry, Northern Ireland.

About the Author